建筑广场系列丛书 No.57

BREATHE
能源意识与可持续公共空间
Energy-conscious and Sustainable Public Spaces

汉英对照
（韩语版第373期）

韩国C3出版公社 | 编

安雪花 于风军 焦明 孙探春 杜丹 徐雨晨 | 译

大连理工大学出版社

4

004 Echavacoiz城市电梯 _ AH Asociados

010 埃尔纳尼的城市电梯与人行天桥 _ VAUMM

016 巴拉卡电梯 _ Architecture Project

22 能源意识与可持续公共空间

022 能源意识与可持续公共空间 _ Julian Lindley

028 柔和住宅 _ Kennedy & Violich Architecture, Ltd.

034 拜耳生态商业建筑 _ Loeb Capote Arquitetura e Urbanismo

040 亚利桑那州立大学卫生服务楼 _ Lake|Flato Architects

048 大卫与露西尔·帕克基金会总部 _ EHDD

060 ECCO酒店与会议中心 _ Dissing+Weitling Architecture

072 北京第四中学房山校区 _ OPEN Architecture

088 日压端子工厂 _ Ryuichi Ashizawa Architect & Associates

102 Le Clos des Fées村 _ Mutabilis Paysage + CoBe Architecture

120 2015世博会：绿色实验室

120 *2015世博会：绿色实验室* _ *Marco Atzori*

134 生命之树 _ Marco Balich + Studio Giò Forma

136 意大利馆 _ Nemesi & Partners

142 翅膀 _ Studio Libeskind   143 万科馆 _ Studio Libeskind

144 Copagri馆 _ EMBT   146 奥地利馆 _ team.breathe.austria

152 巴西馆 _ Studio Arthur Casas + Atelier Marko Brajovic

156 英国馆 _ Wolfgang Buttress   160 韩国馆 _ Archiban   164 法国馆 _ X-TU Architects

168 中国馆 _ Tsinghua University + Studio Link-Arc

172 俄罗斯馆 _ Speech Tchoban & Kuznetsov   173 巴林国馆 _ Studio Anne Holtrop

174 阿联酋馆 _ Foster + Partners   175 德国馆 _ Schmidhuber

176 拯救儿童馆 _ Argot ou La Maison Mobile

178 慢食馆 _ Herzog & de Meuron

4
004 Urban Elevator in Echavacoiz _ AH Asociados
010 Urban Elevator and Pedestrian Bridge in Hernani _ VAUMM
018 Barrakka Lift _ Architecture Project

22 Energy-conscious and Sustainable Public Spaces

022 *Energy-conscious and Sustainable Public Spaces _ Julian Lindley*
028 Soft House _ Kennedy & Violich Architecture, Ltd.
034 Bayer-Eco Commercial Building _ Loeb Capote Arquitetura e Urbanismo
040 Arizona State University Health Services Building _ Lake|Flato Architects
048 The David and Lucile Packard Foundation Headquarters _ EHDD
060 ECCO Hotel and Conference Center _ Dissing+Weitling Architecture
072 Beijing No.4 High School Fangshan Campus _ OPEN Architecture
088 Factory on the Earth _ Ryuichi Ashizawa Architect & Associates
102 Le Clos des Fées Village _ Mutabilis Paysage + CoBe Architecture

120 EXPO 2015 : The Green Laboratory

120 *EXPO 2015: The Green Laboratory _ Marco Atzori*
134 Tree of Life _ Marco Balich + Studio Giò Forma
136 Italy Pavilion _ Nemesi & Partners
142 The Wings _ Studio Libeskind   143 Vanke Pavilion _ Studio Libeskind
144 Copagri Pavilion _ EMBT   146 Austria _ team.breathe.austria
152 Brazil _ Studio Arthur Casas + Atelier Marko Brajovic
156 UK _ Wolfgang Buttress   160 Korea _ Archiban   164 France _ X-TU Architects
168 China _ Tsinghua University + Studio Link-Arc
172 Russia _ Speech Tchoban & Kuznetsov   173 Bahrain _ Studio Anne Holtrop
174 UAE _ Foster + Partners   175 Germany _ Schmidhuber
176 Save the Children Village _ Argot ou La Maison Mobile
178 Slow Food Pavilion _ Herzog & de Meuron

Echavacoiz城市电梯 _AH Asociados

在通过Echavacoiz Norte区的行人移动性方面的研究项目里，三个存在交通不便、城市融合等历史问题的重要区域得到了重视，这些问题可以通过现代机械系统解决。其中一个重要区域需要解决人行坡道与楼梯危险的问题，这些通道是用来调节30m高差的。Urdanoz Group区的居民也通过这些通道走到位置较高的另一侧，那里有周边人行道和Echavacoiz Norte区。城市电梯的设计用两架步行天桥和一部电梯就解决了当地的交通问题，让Echavacoiz从一个只有虚名的区域变成一个拥有自己城市版图的名副其实的城市，让人行天桥和河岸公园在能够互相连接的同时，可以在未来接通快速铁路，让人行道也可以像汽车道一样，能够连接两个高度不同的区域，而这样的人行天桥也同时成为城市的一道独特靓丽的风景。

城市电梯的建筑方案着重强调建筑风格的简洁性，避免过度的结构形式主义。连续横梁构成了人行天桥的基本形状，并支撑着人行天桥的路面。路面由金属薄板制作而成，褶曲的钢板覆盖横梁表面和塔楼外侧，在视觉上实现整体的连贯性，增强城市特色。人行天桥的水平部分在与塔楼的相接处转化成垂直设计，呼应了钢结构构件形式和钢质表皮的设计。不对称的人行天桥可以保护行人免受大风的影响，同时可以让人们欣赏到全新的景观；垂直的支柱也能够牢固地支撑人行天桥。面向景观的电梯出口处只采用了很少的几种建筑材料。

Urban Elevator in Echavacoiz

In study on pedestrian Mobility in "Echavacoiz Norte", three critical areas with historical accessibility and urban integration problems were detected and could be solved by implementing mechanical systems. One of these three critical areas was to resolve the precarious pedestrian ramp access and stairs which overcome thirty meters height difference between levels. These accesses were also used by neighbors of "Urdanoz Group" to reach the elevated area where there was a perimeter walkway and the neighborhood of "Echavacoiz Norte". This project was intended to solve current accessibility problems through two footbridges and a lift, which turned an urban reference of the integration of Echavacoiz into the city and into an object sensible to its own urban landscape. This has been possible by linking the upper pathway with the river park and with the future neighborhood of the AVE. This made

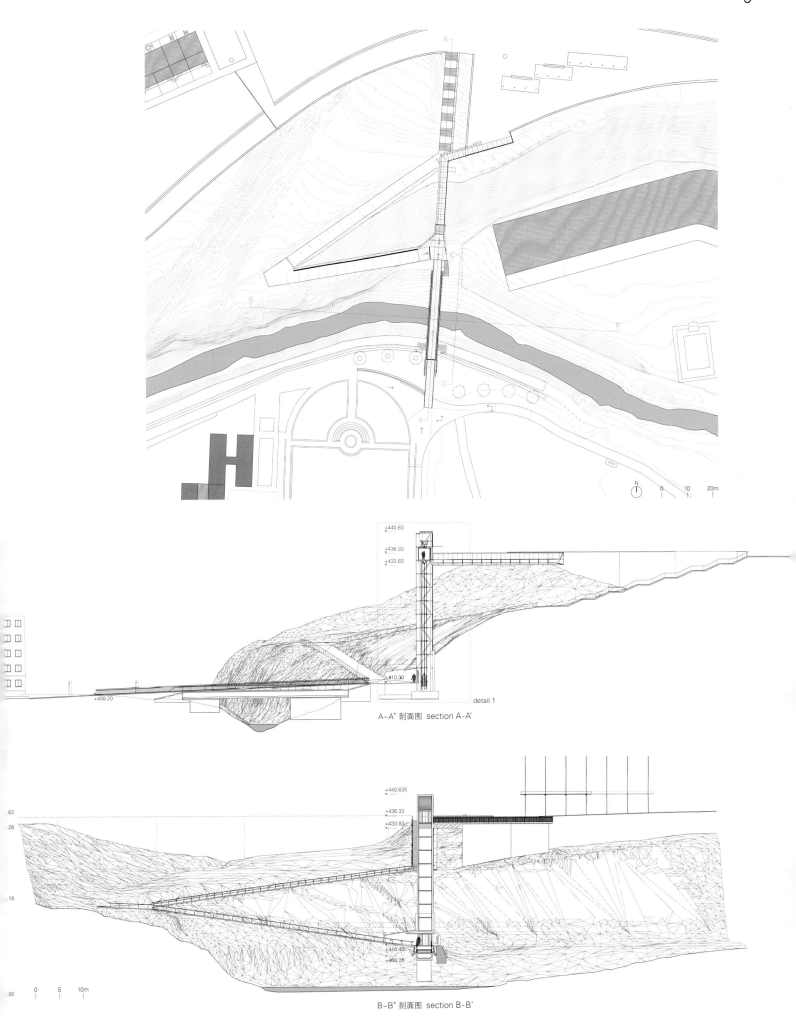

A-A' 剖面图 section A-A'

B-B' 剖面图 section B-B'

it possible to introduce new pedestrian and cyclist roads between the two urban levels and implement an architectural element that turned the panoramic footbridge and the panoramic tower into one.

The project enhances the simplicity of each element, avoiding any excess of constructive formalism. The basic shape of the footbridge is formed by a continued beam from which the supports of the footbridge pavement are born. This pavement has been made of sheet metal plates. The exterior of the beam and the lateral levels of the tower are also covered by a folded steel plate to get visual continuity to enhance the urban character of an element that emerges from the hill and is supported by the head of the footbridge. The horizontal part of the bridge turns into vertical where it meets the tower, in such a way that the format of steel structure element and steel skin is repeated. The asymmetry of the footbridge protects the pedestrians from wind and let them see a new territory whilst the vertical element is robust and strong in its lateral levels. The landings are open to the landscape with the minimum expression of materials.

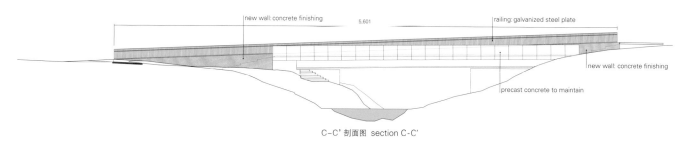

C-C' 剖面图 section C-C'

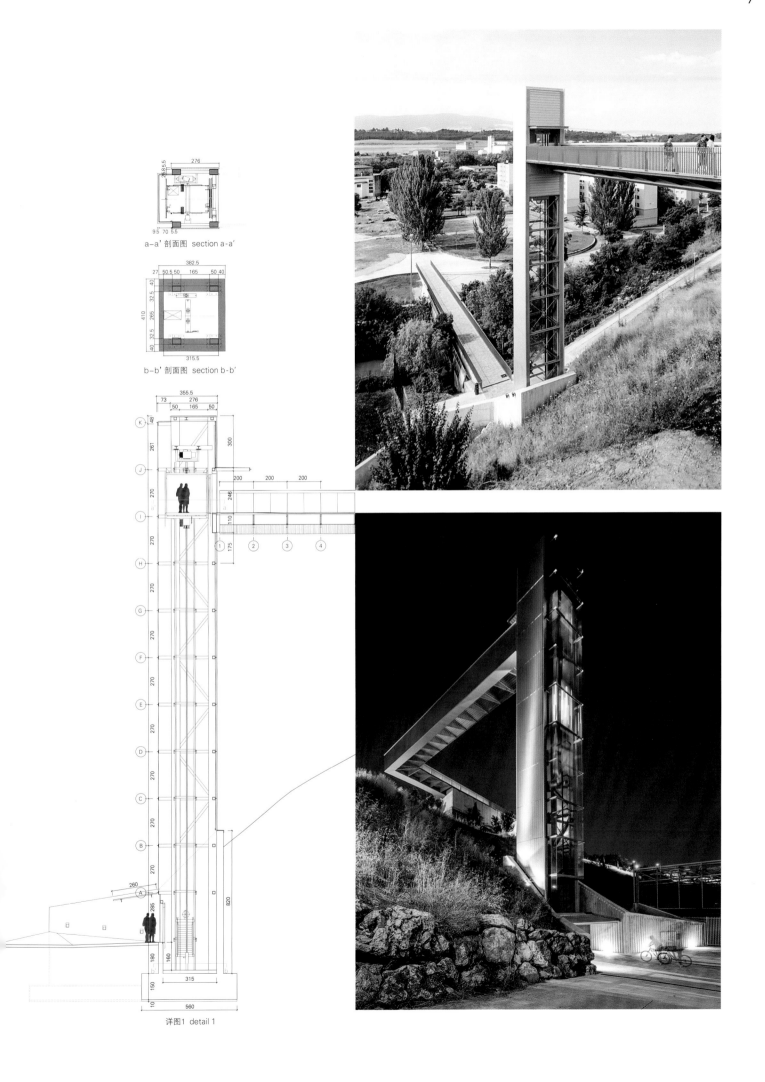

a-a' 剖面图 section a-a'

b-b' 剖面图 section b-b'

详图1 detail 1

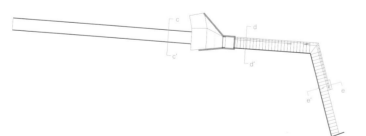

1. horizontal galvanized steel plate, thick = 10mm height = 80mm
2. galvanized steel plate, thick = 8mm height = 60mm every 10cm
3. galvanized steel railing diameter 5cm
4. Galvanized steel support every 200cm, thick = 12mm
5. horizontal galvanized steel plate, thick = 10mm height = 60mm
6. galvanized steel anchor plate, thick = 8mm
7. galvanized steel plate finish, thick = 3mm
8. embedded lighting fixture type, embedded under the gangway finish
9. stainless steel profile 50.50.3 welded to a base plate, height = 120mm thick = 3mm, fixed with stainless steel screw and neoprene separator
10. folded aluminum sheet in book shape FALKIT type "Sierra de Cazorla" Serie 200 ref.103102 or similar
11. steel box beam, thick = 10mm, dimensions according structure plans
12. longitudinal plate for brackets joins 110x10mm
13. longitudinal galvanized steel stiffener, thick = 15mm
14. cross galvanized steel stiffener every 200cm aligned with the bracket support
15. stainless steel rectangular profile 70.40.3, as support for checker plate with same material chocks, even neoprene joint
16. longitudinal steel stiffener, thick = 15mm height = 150mm
17. expanded polystyrene, thick = 5cm
18. galvanized steel "L" shaped profile

项目名称：Urban Elevator in Echavacoiz
地点：Echavacoiz neighborhood, Pamplona, Spain
建筑师：Miguel A. Alonso del Val, Rufino J. Hernández Minguillón, Marcos Escartín Miguel, Mikel Zabalza Zamarbide _ AH Asociados
项目经理：Patricia Biain Ugarte
项目团队：Esperanza Marrodán Ciordia, María José, Alonso Pérez, Xabier Eskisabel Azanza
结构工程师：Eduardo Ozcoidi Echarren
设施：Javier Gironés Navarlaz
S & H：Michel Aldaz García-Mina
客户：Strategic Projects Area, Pamplona City Council
设计时间：2012.1 / 施工时间：2012.10 / 竣工时间：2013.7
摄影师：©Jesús Lázaro Izquierdo (courtesy of the architect)

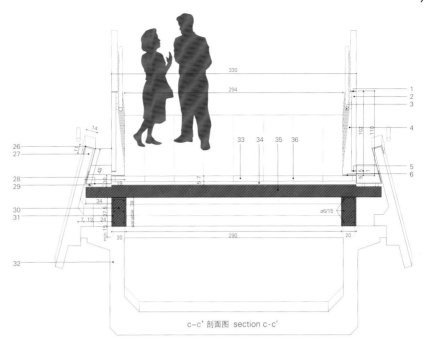

c-c' 剖面图 section c-c'

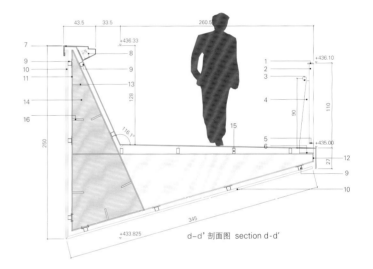

d-d' 剖面图 section d-d'

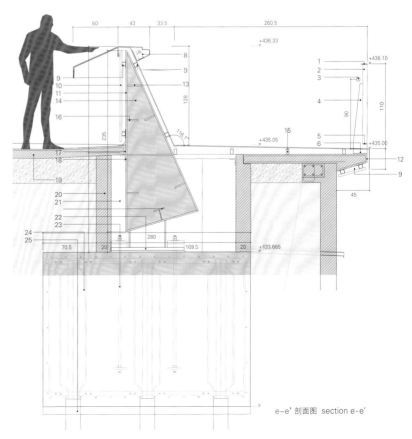

e-e' 剖面图 section e-e'

19. concrete slab similar to existent
    with steel reinforcement 8/15
20. concrete wall 20cm with double steel reinforcement 6/20
21. lift ditch for regular inspection
22. steel plate 1000.1000.35 over a high resistance mortar layer
23. 12 steel anchor bolts #32-6.8
24. reinforced concrete foundation 280x600x200cm
25. concrete micro piles ø200mm
26. auction galvanized steel perimeter sill prefabricated
    supported in a channel and steel overlapping, thick = 5mm
27. sill part of existing prefabricated concrete
28. folded sheet steel galvanized, thick = 5mm, by way of canal
    with longitudinal slope of 3.7%
29. omega galvanized steel channel substructure 40,40,20mm
    as steel and screwed a precast slab
30. beam of concrete, thick = 20cm, support for precast slab,
    anchored beam bridge through holes filled resin hilti re = -500
31. double mesh ø6/15
32. prefabricated girder bridge calculated to overload 1000kg/m²
33. tile floor type, color ash mass, dim. 60x40x7cm. 40x40x7cm
34. grip mortar layer M40, thick = 4cm
35. concrete slab type T7, thick = 13+3cm
36. fill mortar board

埃尔纳尼的城市电梯与人行天桥 _VAUMM

在城市可持续发展的自然变革过程中，景观的可接近性显得尤为重要，城市基础设施的创建要更加人性化，每一个公民都应该有机会去接近它们。本案项目中的城市电梯将Latsunbe-Berri区和Urbieta大街连接起来，同时，这个公共设施也是将老城区和新城区连接起来的纽带，不只是地域的联系，它的范围包含城市发展的各个方面。若把范围缩小一点，该电梯还将运动中心和康复中心连接起来，满足了人们日常生活中进入这些重要公共设施的需要。

人行天桥的设计深受运动中心的位置决定和影响，桥的背面是一个绿地斜坡，从景观设计的角度来看，该坡道通往远处的大山。而人行天桥巧妙地运用了这样的双重景观，一面是封闭的，另一面是开放的。不管是电梯还是天桥的设计，都与人们紧紧相连，人们可以乘电梯通往人行天桥，穿过天桥便到达了一个树林，再往前，视野又会开阔，可以看到更远处的景象。

海拔较低的塔楼可以实现将电梯入口修建在树林中，彰显电梯的城市化特征。电梯的中间站既解决了进入体育中心Pelota赛场的问题，也作为一个悬臂梁支撑着整个人行天桥。Urbieta大街上的人行天桥被拓宽出一个充足的空间供行人通过。

从结构的角度来看，塔楼的支撑能够使人行天桥更加厚重，也就让人行天桥能够有足够宽的路面通过原有的挡土墙来支撑。而正是由于这些结构要求，人行天桥将会创新性地打破传统模式，形成独特的雕塑艺术表达语言。

利用这种表达语言，整座塔楼和人行天桥得以改造其所在的城市环境。黑色的基调渲染了抽象的艺术氛围，形成了强烈的对比效果：一面是透过玻璃看到的璀璨的城市景观，另一面则是一片漆黑，似乎是在尝试与宽敞的体育中心进行对话。

Urban Elevator and Pedestrian Bridge in Hernani

In the natural evolution of a city towards sustainability, accessibility may be seen as a key element, capable of providing opportunities for all population segments of citizenship. The elevator connects Latsunbe-Berri district with Urbieta Street,

项目名称：Urban Elevator and Pedestrian Bridge in Hernani
地点：Hernani, Gipuzkoa, Spain
建筑师：VAUMM
结构工程师：Raul Lechuga Durán
发起者：City of Hernani
项目团队：Iñigo garcia odiaga, Jon muniategiandikoetxea markiegi, Marta álvarez pastor, Javier ubillos pernaut, Tomás valenciano Tamayo
结构设计：Raul Lechuga Durán
质量控制与管理：Julen Rozas Elizalde
健康与安全：Bategin
用地面积：355m² / 建筑面积：160m² / 有效楼层面积：155m²
竞赛时间：2014.6 / 设计时间：2014.7 / 施工时间：2014.11—2015.4
摄影师：©Aitor Ortiz (courtesy of the architect)

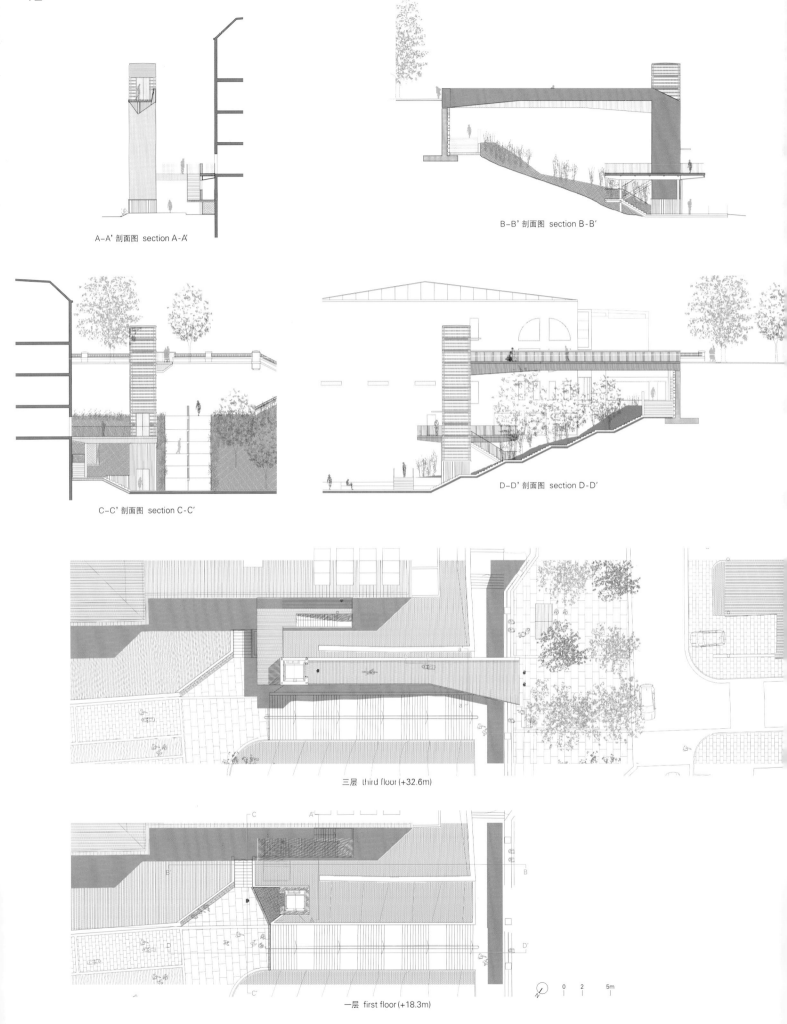

A-A' 剖面图 section A-A'

B-B' 剖面图 section B-B'

C-C' 剖面图 section C-C'

D-D' 剖面图 section D-D'

三层 third floor (+32.6m)

一层 first floor (+18.3m)

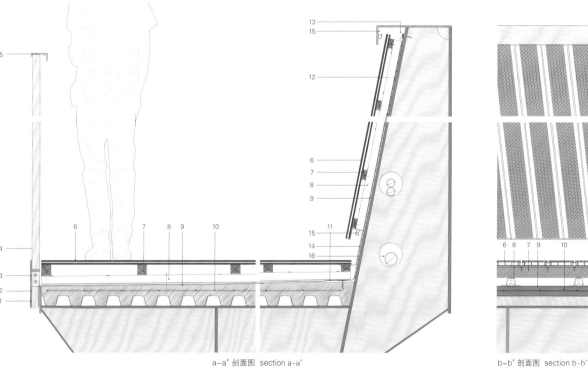

a-a' 剖面图 section a-a'   b-b' 剖面图 section b-b'

1. outer rim structure 2. cross deck frame to axle, structure 3. bottom rail frame fixed to transverse plate 4. vertical deck railing, forges plane defined 5. insulation behind 6. engineered wood made from recycled wood with polymeric resins, profiles 138mm wide, 23mm thick and variable length, 5mm space between profiles, concealed fixing system using steel clips and fasteners stainless 7. technological wood battens width 50mm, distance wheelbase 40cm 8. reinforced polypropylene continuous profile type 9. EPDM waterproofing membrane type 10. forged steel decking reinforced with mesh distribution ø5 11. sure seal reinforcement 12. contact adhesive type adhesive for bonding sure seal sheets adhere to various media 13. sure seal termination bar 14. matte stainless steel profile 3mm thick 15. LED luminaire, lamp model fine leds strip 16. holes in the structure stiffeners

but also blurs an important topographic difference of level, that until now, it caused the fracture of the center of the town of Hernani. This way, the infrastructure serves as a link between the old town and new town developments on a wide scale, while on a closer scale, serves as a link between the Sports Center and the Health Center, providing accessibility to these important public facilities in everyday life.

The construction site has been strongly influenced by the presence of the Sports Center, which is opposed to a green slope that connects, from a landscape point of view, with the distant mountains. The pedestrian bridge uses this double situation, closed on one side and open on the contrary, relating the users, from the elevator inside, and while passing through the pedestrian bridge, with the nearby woodland and the faraway landscape.

Functionally, the emergence of concrete tower on the lower level, let us build an access space that by incorporating a wood bank, articulates the elevator with the existing urbanization. The elevator's intermediate stop, which solves the accessibility to the Pelota Court of the Sports Center, also functions as a large cantilever covered space. On arrival at Urbieta Street, the widening of the pedestrian bridge also builds a generous area, inviting passersby to approach.

From a structural standpoint, the pedestrian bridge's structural section gets thicker in the support of the tower and widens the surface to rest on the existing retaining wall. Due to these structural requirements, the geometry of the pedestrian bridge will progressively change its section, building a faceted and angular shape, about to become sculptural language. Through this kind of language, the whole tower and pedestrian bridge propose to organize the urban environment in which they are located. The black color of the set gives abstraction to the piece, which offers two opposing faces, one of the lightness to the landscape, characterized by the use of glass, or thin banisters, and the other rotundity of the monolithic black piece, which tries to dialogue with the huge scale of the Sports Center.

巴拉卡电梯_Architecture Project

使港口与城市形成纵向连通,并非推陈出新、标新立异。建于1905年的巴拉卡电梯有效地连接了格兰德港口和马耳他首都瓦莱塔,这部电梯运行了将近70年,已然成为马耳他现代化的地标建筑。

港口新兴商业的出现,使其成为众多豪华游轮青睐的终点站,由此带动了马耳他经济的复苏。成千上万的旅客的到来,使得一个行之有效的方案出台迫在眉睫,即能够使游客从港口快速进入位于港口水平线之上的市区。加之当地人想让马耳他成为受欢迎的休闲胜地,于是,便有了动力修建巴拉卡电梯。

电梯的设计考虑了多方面因素:从格兰德港口到瓦莱塔之间的客流;电梯依附的历史古墙;旅客的体验,即水平移动时经过Lascaris Ditch,纵向移动时可以欣赏到堡垒和港口风景,然后到达巴拉卡公园;整个设计建造体系完全符合并遵照联合国教科文组织关于世界文化遗产的规定。

电梯平面的几何特点与城墙的角度轮廓相呼应,高度和比例参考了经典的多立克柱设计。垂直的部分是搭乘入口;电梯塔楼被波浪状的铝质表皮覆盖,体现了格兰德港口的工业化遗产文化;台阶的材质是石头和水泥,混合着些许石灰岩。

金属网的透明度在电梯的实心结构和镂空结构间不断变化,其光影效果营造了如同丝质般的表面。温和的青铜色和柔和的反射特点随着阳光照射的强度和角度的变化,形成了多彩变幻的曼妙光影。而在人们记忆深处,金属网似乎象征着与古老的钢材的联系,呈现出历史穿越未来的魔幻视觉效果,像Jules Verne这样的作家就曾这样描述——当代设计与老电梯的完美结合。

Barrakka Lift

The need for a vertical link between the harbour and the city is not something new. Built in 1905, the Barrakka Lift efficiently transported the activity generated by the Grand Harbour to Malta's Capital, Valletta. Becoming a symbol of Malta's adoption of modernity, the lift remained operational for almost seventy years.

The emergence of new activities within the harbour, spearheaded by Malta's increasing popularity as a cruise liner destination, signal the reversal of decline. The arrival of thousands of tourists has reactivated the need for an efficient solution to transport pedestrians from the harbour to the city above. This, coupled with the desire to make the harbour a popular leisure destination for the local population was the primary objective behind the rival of The Barrakka Lift.

The design of the lift took into consideration a number of factors; pedestrian flow between the Grand Harbour and Valletta; the context of its historical surroundings; the user experience – from the horizontal movement through Lascaris Ditch, vertically through the two lifts with views of the bastions and the harbour, to the arrival in Upper Barrakka Gardens; and the analysis of a system that would be completely reversible and respectful of the UNESCO World Heritage site.

Its shape and form relate to the outline of the fortifications of Valletta, its height and proportions relate to those of the classical Doric column. The vertical components suggest the threshold they sit in; the lift-tower wrapped in an anodized aluminum mesh reflects the industrial heritage of the Grand Harbour; the stair-structure is built out of stone and concrete, blending with the limestone bastions.

The varied translucency of the mesh plays between the solid and open parts of the structure, manipulating light and shadow producing the veil-like quality of skin. Its mild bronze color, and subtle reflective properties allow for a delicate change in color depending on the intensity and angle of the sun. The metallic mesh, a token psychological link to the steel of the old, conjures historic visions of future, typical of writers such as Jules Verne – a contemporary with the old lift.

项目名称：Barrakka Lift
地点：Lascaris Ditch, Valletta, Malta
建筑师：Architecture Project
照明设计：Franck Franjou
客户：Grand Harbour Regeneration Corporation plc
可容纳人数：800/hour
造价：EUR 2,000,000
施工时间：2009—2013
摄影师：
©Bettina Hutschektos (courtesy of the architect) - p.20
©Luis Rodríguez López (courtesy of the architect) - p.17, 18[left], 19
©Sean Mallia (courtesy of the architect) - p.16, 18[right], 21

南立面 south elevation

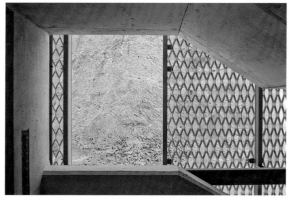

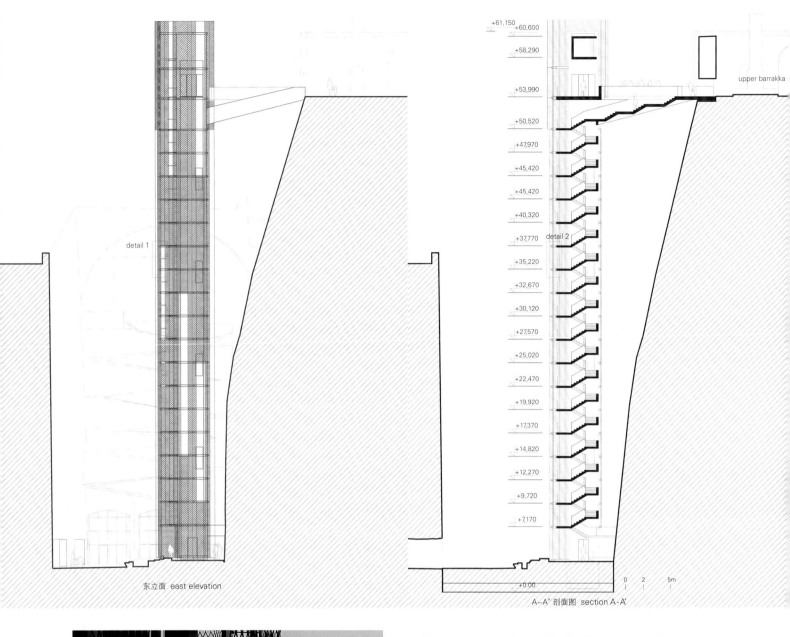

东立面 east elevation

A-A' 剖面图 section A-A'

1. architectural screen
2. height of railing has to comply with KMPD and H&S regulations and has to match with the height of concrete railing around void
3. galvanized steel section
4. galvanized steel flat bars
5. galvanized steel L-section connection of railing to secondary structure
6. connection of railing to the concrete lift shaft
7. secondary structure
8. spacer
9. drip detail

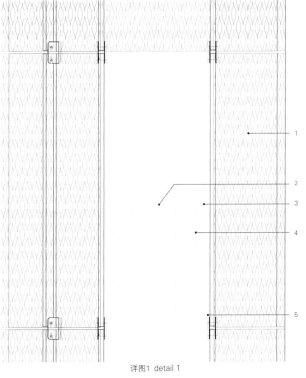

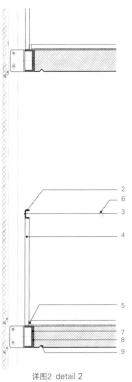

详图1 detail 1　　详图2 detail 2

能源意识与可持续公共空间

　　可持续发展以及建筑设计已被讨论多年,然而,由于需要时间去设想、补充和应用,所以直至今日我们才可以用现有的建筑,批判地检验建筑师对环境问题的考虑。这篇文章介绍了可持续发展的两个重要方面,如果要做区分的话,就是能源管理以及公共/私人空间。事后诸葛亮也总是有用的,而我们人类这个群体也正在慢慢了解,但不是我们几年前就知道的可持续发展的重要性,而是什么才是真正好的建筑。世界各地的典型建筑都是很好的例子,但同时产生了更多关于我们到底想要什么样的成果的讨论。每一个成功的范例都有各不相同的方法,而且不只局限在对建筑本身的建材来源、地点选择、建筑构造、建筑用途和建筑维护负责,也对资源、能源和水源负责。这所有的一切,都属于可持续发展能力的范畴。以这些典型建筑为例,我们要用更宏观的生命意义来重新定义我们的责任。值得注意的是,我们在能源规划时对地区环境要有不同的理解,不管是城市还是乡村。相同的能源规划有可能在一个地区适用,

Sustainability and Architecture have been discussed for many years. However, due to the nature of time to conceive, implement and use it is only now that we can critically evaluate architects' response to environmental questions through actual buildings. This article links two key, if differing, aspects of sustainability, energy management and public/private spaces. Hindsight is always useful and we are slowly, as a collective, understanding not the importance of sustainability, we acknowledged that many years ago, but what good practice actually is. These examples from across the globe are fine examples but also generate more questions about what we are trying to achieve. Each takes a differing approach to materials, energy and water taking responsibility beyond just the structure to material sourcing, site clearances, construction, use and afterlife, these are now all part of the sustainable equation. This, as demonstrated by the examples, creates a broader "whole life" understanding of our responsibilities. Noteworthy is the understanding of the local environment either urban or rural in energy plans. What works for one site will not for

柔和住宅_Soft House/Kennedy&Violich Architecture, Ltd.
拜耳生态商业建筑_Bayer-Eco Commercial Building/Loeb Capote Arquitetura e Urbanismo
亚利桑那州立大学卫生服务楼_Arizona State University Health Services Building/Lake|Flato Architects
大卫与露西尔·帕克基金会总部_The David and Lucile Packard Foundation Headquarters/EHDD
ECCO酒店与会议中心_ECCO Hotel and Conference Center/Dissing+Weitling Architecture
北京第四中学房山校区_Beijing No.4 High School Fangshan Campus/Open Architecture
日压端子工厂_Factory on the Earth/Ryuichi Ashizawa Architect & Associates
Clos des Fées村_Le Clos des Fées Village/Mutabilis Paysage + CoBe Architecture

能源意识与可持续公共空间_Energy-conscious and Sustainable Public Spaces/Julian Lindley

而在另一个地区就不适用，要因地制宜。所以最重要的是，我们要从这些优秀的建筑中汲取好的方法和灵感。经典建筑之所以成为经典，是因为其更多地解读可能性，并能够引发更多的讨论。这个价值可能不在建筑本身，而是建筑师们在其作品中投入的思想。许多建筑师注重人类的生活质量，重视健康和学业，而忽略了要求，结果使得在人类活动范围内不合理的结构和空间导致生活质量提升缓慢。我们汲取的教训之一便是，所有有可持续发展能力的成功建筑空间设计，最初都花费了大量的时间去了解住户们的需求。综合来看，你会发现各色住户不同要求的共同特征就是呼唤创造力，而你会在本篇文章中看到这样的例子。所有的建筑师、策划者和设计师都应该在观察他人的设计的过程中形成自己的见解、质疑和目标。同时，我们要感谢那些做出贡献的建筑师们，感谢他们的设计在不同程度上增加了这篇文章的讨论性。

another. It is essential that we take stock of approaches and ideas, reflecting on what has been achieved in these examples and what else we can do. What makes a good space is far more open to interpretation and the examples form a polemic for discussion. The value is possibly not in the buildings but the ambitions which the architects want from their work. Many put human qualities such as health and learning at the forefront of specifications resulting in structures and spaces on a human scale inviting slowness from a human quality into function. One lesson to be learnt is that good results, sustainable buildings and spaces, come from the initial understanding of the needs of all stakeholders and the time spent understanding these. Mix with this that unclassifiable quality called creativity and you have the examples presented in this article. It is essential for architects, planners and designers to see the work of others and form their own debates, questions and aspirations from this article, and we thank and applaud the architects who contributed, in a small way additions to these discussions.

法国帕吕埃尔Clos des Fées村
Le Clos des Fées Village in Paluel, France

评论建筑物或公共空间的设计通常是有争议的，因为我们不知道要用怎样的标准来界定成功。而在准备写这篇介绍可持续发展的公共空间的文章时，这个问题就成为一个切入点。降低能源消耗或提高能源利用效率的建筑，就是符合生态效益的建筑，那我们其实可以比照固定的参数和数据来找出符合条件的建筑。我们可以寻找、监督、呼吁新兴的技术用在建筑上来减少对生态环境的影响。接下来我会用具体的建筑案例，如，Dissing + Weitling建筑公司设计的利用地热能的ECCO酒店与会议中心，Kennedy & Violich建筑事务所设计的柔和住宅和Mutabilis Paysage + CoBe建筑事务所设计的Clos des Fées村，来说明"清洁能源"如何取代以煤炭和石化能源为主的能源。

这里需要假设建筑中需要和能使用的能源已经提前决定好了，建筑师只需要用一种有效率的方式去分配这些能源。那么生态效益如何确定呢？也就是说，建筑师要先考虑能源要怎么使用，并在建筑过程中不断提出不同的方案。这样做对新建筑项目的好处就是建筑师、规划师和工程师能够完成一个全面的能源计划，提前了解能源的需求，并将能源需求同调节温度、水量消耗等一一对应。在大卫与露西尔·帕克基金会总部项目上，EHDD建筑事务所将重点放在了需求上而不是消耗上，从而使能源需求量减少了65%。ECCO酒店与会议中心是最大限度地利用空间的一个很好的例子，这也是能源设计的一部分。柔和住宅也是很好的例子，它位于城市网格结构中，更需要创新的方式，充分利用有限的能源。

可持续的公共空间是一个更加有争议的课题，因为我们需要同时理解可持续发展和公共空间两个概念，既要进行无形的讨论，又要进行定性评估。功能性及技术说明也很快被定义。能源需求、环境破坏和资源消耗也包含在内，值得注意的是这些典型建筑强调的是材料与能源从消耗到重建和再利用的过程。

随着建筑方法的不断增多，在建筑设计阶段会有很多新的问题，例如，我们怎么能让建筑在可持续发展方面有更多的成绩？EHDD建

Reviewing Architecture and/or Public Spaces is problematic in that what criteria should we use when evaluating success? This question was the starting point when asked to write this introduction to Sustainable Public Spaces. Buildings which reduce energy usage or do more with less, this is eco-efficiency and to some degree this can be measured given the right parameters and supplied data to evaluate this. We can look for, or review and hail the use of new or emergent technology in buildings to reduce ecological impacts. Here, the next few pages, give good examples of how "green technologies" are replacing carbon or petrochemical based energy sources with geothermal ECCO Hotel and Conference Center by Dissing + Weitling Architecture, Soft House by Kennedy & Violich Architecture, Ltd. and Le Clos des Fées Village by Mutabilis Paysage + CoBe Architecture.

But this makes an assumption that the need and use of energy have been pre-determined and it is only the architect's role to deliver it in an efficient way. What about eco-effectiveness? That is, questioning what the energy will be used for and therefore providing different approaches through built spaces. Here there are advantages with new building projects as architects, planners and engineers can create a holistic energy plan which reviews requirements and connects energy need with other environmental issues such as heating/cooling and water consumption. The David and Lucile Packard Foundation Headquarters, EHDD achieved a 65% reduction in energy need, the emphasis being placed on need not use. The ECCO Hotel and Conference Center is a good example of utilising optimum organisation of space as part of an energy plan. Also to be applauded is the Soft House which is situated in existing urban networks and structures and required a creative approach to energy within tight specifications. Sustainable public spaces are far more problematic in that they demand an understanding of both Sustainability and Public Spaces, both intangible debatable and qualitative in assessment. Functionality this can be defined quickly along with the more technical specifications. Included here could be energy requirements. Environmental damage and resource depletion could also be included and it is worthy to note the examples which highlight the material/energy flows from demolition through rebuild and use.

The expansive approach allows for new questions to be asked at the design stage, such as how can we make our buildings do more? EHDD includes a transport management plan to

美国亚利桑那州立大学卫生服务楼
Arizona State University Health Services Building, USA

筑事务所设计了一个交通管理计划来减少碳足迹。这本身当然是有益的，但有趣的是它需要人类改变自己的一些行为习惯来减少对生态环境的影响，这就引发了一个有趣的大众观念和公众形象的问题。建筑或公共空间是不是应该炫耀它们的可持续发展能力？针对达成可持续发展能力的探讨点是我们现在仍然处于培养可持续发展意识的教育阶段，我们可以利用城市中的建筑来定义可持续发展的特征，从而呼吁体制机制的改革，就像柔和住宅，在自身外部视觉效果上显示出它的节能效率。然而，这不是要求所有的有可持续发展特征的建筑都要明显具备这一特征。所有的建筑都需要具有自身可持续发展的能力，但可以隐藏在它们的内部装修设计上。建筑师们需要在设计过程中说明建筑中所具有的可持续发展能力。通常，可持续发展能力是隐藏在建筑中的。Open建筑事务所设计的北京第四中学房山校区就是一个很好的高效率利用能源的典范。

能源的使用、材料的选取、水的管理方式是减少生态破坏或资源浪费的关键因素，那么关于另一个可持续发展的方面——人类健康（Okala 2004 & 2013）或者生活质量（Fuad-Luke 2009）又该如何体现呢？如果我们没有量化指标依照的话，我们如何判定我们做的是可持续发展设计呢？以上的说法似乎不管对建筑还是对这篇导言来说，都是一个很好的开端。用户（那些使用或受影响的人）或者实行者（积极影响决定的规划师和建筑师或者说使用者）对他们的建筑有什么样的需求？社会或文化需求方面大致都被分到了可持续发展能力的第三方面，然而可惜的是，对这些方面的定义却不够明了。文章内容包括公共空间，那么怎样才能让人们有更好的机会同时产生一种生活的满足感和社会责任感？可能一个慢节奏的空间会让住户在这个速度中享受的同时培养出效率意识和社会意识。这不就是我们正在寻找的吗？LakelFlato建筑师事务所设计的亚利桑那州立大学卫生服务楼巧妙地证明了这一观点。

他们这样描述他们的一个目标：

reduce the carbon footprint. This in itself is beneficial but it is interesting in that it demands a change in human behaviour to achieve a lowering of ecological impacts. This raises an interesting question of perception or public image. Should buildings or spaces flaunt their sustainable credentials? The argument for an overt approach to sustainability is that we are still at the stage of sustainable education, raising the awareness of the issues and the mechanisms by which we can address them set in an existing urban setting, the Soft House celebrates its energy credentials in the visual language used in the exterior of the building. However this is not to demand that all sustainable buildings have to be overt in their message. All buildings need to be sustainable and the sustainable credentials covert in their implementation. Sustainable criteria are another layer of requirements within a building's specification which architects need to respond to when proposing buildings. This most of the time sustainability can be covert. Beijing No.4 High School Fangshan Campus by Open Architecture is a good building and space which is also efficient in its use of energy.

Addressing energy, material flows and water management is key to reducing ecological damage and resource deple-tion but what about another aspect of Sustainability; Human Health (Okala 2004 & 2013) or even Well-being? (Fuad-Luke 2009) But how do we judge sustainability if we do not have any quantitative scales to work against? The above observation seems the perfect starting point for architecture and certainly this introduction. What do the stakeholders (those who use and/or are affected by) or actors (those who actively influence decisions, planners, and architects and dare I say it, users) need of the buildings they construct? Social or cultural aspects are commonly grouped as the third field of sustainability, but sadly ill-defined. The article includes public spaces, and what better opportunity is there to bring people together to engender a sense of well-being and community spirit? Maybe a space which is "slow" giving users purpose to use admire or enjoy at a speed which will foster both efficiency and community. Is this what we are seeking? An interesting standpoint is provided by Arizona State University Health Services Building, Lake|Flato Architects.

They state the following as one of their objectives:
"to transform a sterile inefficient health clinic into a clearly organised and welcoming facility"
The architects continue to consider how the natural environ-

马来西亚柔佛日压端子工厂
Factory on the Earth in Johor, Malaysia

丹麦Tønder ECCO酒店与会议中心
ECCO Hotel and Conference Center in Tønder, Denmark

"将一个毫无生气、效率低下的卫生服务诊所变成一个组织有序、受欢迎的卫生服务中心。"

建筑师还考虑如何综合利用自然环境使校园适于行走，他们也赞成将校园历史遗址作为行人走路的核心区。有趣的是，建筑师会拿出建筑前后的照片对比，设计的目的核心用语言来说明，就是为了用户获得幸福感。许多建筑案例就是创造了一个能让自然、景物和植物同人类交流的建筑空间。许多建筑事务所从环境的角度考虑了空间的独特性，并结合具体知识技术将其应用在建筑中（Ryuichi Ashizawa建筑师及合伙人事务所设计的日压端子工厂，Loeb Capote Arquitetura e Urbanismo设计的拜耳生态商业大楼）。这些例子是令人兴奋的建筑设计挑战，在技术上、视觉上和选址上，都体现了一种国际现代建筑理念的融合。

在大卫与露西尔·帕克基金会总部项目中，我们不能明确定义内部花园到底是一个室内房间还是一个室外庭院，而无论是哪一个，这样的设计都鼓励慢生活节奏，不管是个体的、大众的还是工作的节奏都要慢下来，甚至连结构和材料都给人一种有机的设计感，而非机器制造的简单粗暴。在很多建筑案例中都有的另一种理念是从建筑的功能需求出发进行分析。ECCO酒店与会议中心用建筑最终的功能来决定建筑的设计方案，这样就使建筑的空间分配利用更加合理，更具有投资价值。而这样设计的灵活性正好体现出效率、人员和能源的稳定性，除非要改变建筑的最初功能性。北京第四中学房山校区就是一个建筑功能性明确，通过空间分布的需求而分析出符合建筑目的的设计。与中国其他城市的校园设计相比，北京第四中学房山校区是一个自由组合的多中心的建筑空间，用多重的走廊设计让学生融入大自然。

文章写到现在，我思考了一下从我们提到过的案例分析出来的可持续标准，第一点是能源，接着是人文。而这不可能是完全正确的，因为一座建筑必须从各个方面都满足所有用户的要求。

衡量成功的最终标准就是看建筑师们如何用创新的方法来协调各个方面的需求，既要给用户们提供功能齐全的私人、小众、大众、社会或商业的空间，又要尽可能小地影响自然环境，同时还要保证用户

ment can contribute to a cohesive pedestrian orientated campus. They also nod to the site's history in engaging the historic palm walk as a pedestrian spine. It is an interesting distraction that the architects have offered photographs of the site before and after construction. The language used in the descriptions places people's well-being at the core of the objectives for the site. Many of the examples use the connection between natural, space, vistas and flora to create spaces and buildings on a scale which people relate to. Many of the Architects have considered the uniqueness of space from an environmental perspective incorporating this knowledge into the efficiency of the buildings (Factory on the Earth by Ryuichi Ashizawa Architect&Associates, Bayer-Eco Commercial Building by Loeb Capote Arquitetura e Urbanismo). This provides an exciting challenge of designing, both technically and visually, for site rather than an amalgam of international or current style.

With the David and Lucile Packard Foundation Headquarters, we are not sure if the inner courtyard is a room or exterior space, and whichever it is, it is actively encouraging people to slow down whether it is personal, public or work time. Even an approach to structure and materials can give a sense of the organic rather than the brutality of machine-made. Another approach which is evident in many of the examples is the initial analysis of the activities which form the functional requirements of the buildings. The resultant circular plan of the Ecco Hotel and Conference Center is a direct result of making the use of space efficient and more fluid. This flexibility implies stability of effectiveness, people and energy, if needs change beyond the initial purpose. Beijing No.4 High School Fangshan Campus is a good example with effectiveness of purpose achieved through analysis of the disparate demands on space. The plan is based on a web of needs rather than a hierarchy, a free form with multiple centers, bulging corridors to accommodate students while integrating "natural" into the built spaces, a direct contrast to the design of other schools in China's urban areas.

So far I have considered the metrics by which we review the examples independently, that is energy first and then people. This cannot possibly be correct as architecture has to be an all-embracing challenge, appealing to all stakeholders. Success is ultimately gauged retrospectively by how clever the

巴西圣保罗拜耳生态商业建筑
Bayer-Eco Commercial Building in São Paulo, Brazil

的幸福感。我们要称赞在接下来的篇章中介绍到的建筑，它们用既有效又环保的方式来平衡用户的各方面需求。我们已经谈到了几个例子，这些建筑用完全创新的方式来达到这些目的。建筑的创新可以是技术上的，也可以是视觉上的。我也可以让你自己设定自己的标准来评判成功的标准。如果这样，我就可以结束这篇文章了，或者从"人"的角度重新审视这些建筑案例。但问题是，我们怎么能决定其他人的需求呢？从这里，我们可以用建材的使用和创造力来评价建筑的功能对人需求的满足程度，并引入人类的幸福感、快乐感和便利性等概念。人们对每一个空间或每一座建筑都有不同的需求。如果想要满足这些需求，就要花足够的时间来分析人们的需求，并通过对建筑或空间的功能设定来解决这些问题。我只想问你，你们所有人，不论男女老少，无论什么种族，不管身强体壮还是体弱多病，如果按照原建造目的来使用这些空间，那会是什么样？

Sir Ken Robinson最近解释："你不能解答一个错误的问题。"[1]

建筑师有责任在开始进行创意性设计时就清楚地知道人们需要怎样的建筑。建筑师，像所有好的设计师一样，需要自由挑战最初的构思，并确定人们的真正需求，而不是一味地迎合或误解。这样，建筑师就可以创造出环境优雅、功能齐全并有可持续发展能力的建筑，让人们可以享受私人空间和公共活动。不同于其他的设计类型，建筑师和规划师有更重大的责任，因为他们的作品会在这个世界上保留很长时间，假设我们不规定每座建筑的使用年限。一些建筑会存在好几个世纪，因此它们对生态环境的影响，不管是好是坏，都要比它们的建筑师造成的影响时间长，并会影响到下一代。我在文章的一开始问了一个问题，而在文章的结尾，我也要问一个问题："我们怎么能保证建筑可以一直保持自身的可持续发展性呢？"

architect(s) was(were) in balancing compromises in new and creative ways, providing users, individuals, groups, public, social or business, buildings or spaces which are functional, have little or no impact on the resources of the planet, and are a joy to experience. We admire the work in the following pages for the skill in which they have balanced the demands of stakeholders in effective and beautiful ways. We have reviewed the examples as they have approached the issues above in different and new ways. Creativity can be evident either technically or visually and here I will allow you to decide on your own parameters to objectively evaluate success.
On that I could finish this article and challenge the readers to objectively look at the architectural examples presented from the perspective of "people", but, and here is the real question, how can we define the needs of people? From that we can evaluate the architectural responses and bring in notions such as well-being, joy, effectiveness along with material use and creativity. Each space or building has different requirements. To effectively respond to it, time needs to be spent on analysing need and defining the human problems which should be addressed through a building or space. I only ask that you include people, lots of them, old and young, multi-ethnicity, able-bodied and less so when you consider what it is like to use these spaces for their intended purposes.
Sir Ken Robinson recently explained that "You cannot solve a wrong problem"[1]
Architects have a responsibility at the start of the creative process in clearly understanding what is needed. Architects, like all good designers, need the freedom to challenge the initial brief and determine what the real needs are rather than respond to perceived or misunderstood demands. In that way they will be able to deliver elegant, functional, sustainable buildings which becomes a joy for both private and public activities. Unlike other design disciplines architects and planners have a greater responsibility as their work remains with us longer, assuming we do not start prescribing buildings with a defined lifetime. Some buildings last for centuries, and so the ecological impacts, good or bad, outlive their creators and affect future generations. I started the article with a question and I finish with one "How do we ensure longevity of sustainability within buildings?" Julian Lindley

1. Sir Ken Robinson, RSA Talk, London, June 2015.

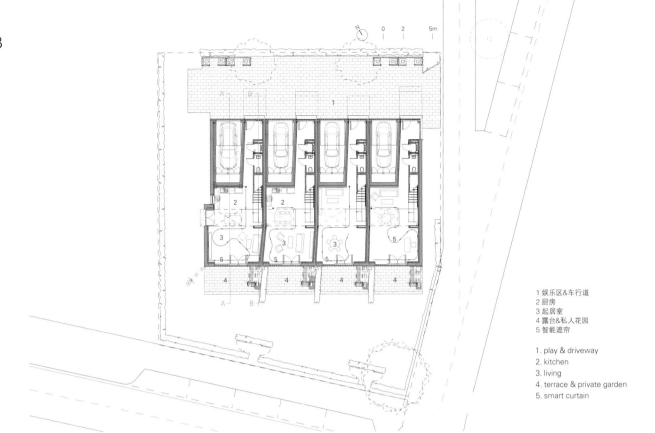

1 娱乐区&车行道
2 厨房
3 起居室
4 露台&私人花园
5 智能遮帘

1. play & driveway
2. kitchen
3. living
4. terrace & private garden
5. smart curtain

柔和住宅

Kennedy & Violich Architecture, Ltd.

德国汉堡的柔和住宅是IBA设计竞赛中获胜的方案，IBA是自1901年以来，德国最著名的建筑传统活动之一。这座建筑于2013年3月竣工，它是可供居住和工作的联排住宅，提供了降低碳排放的新模式以及生态可持续的生活模式，可以人性化地迎合用户需求。柔和住宅展示了如何使住房的基础结构变得"柔和"——灵活的生活理念，碳中性的实木结构，可无线控制的遮阳织物，后者正是这座建筑的独特之处。柔和住宅用重新定义的"软"和"硬"的材料，建筑的完美融合，可移动的织物遮阳帘和清洁能源基础设施，改变了德国被动式能源房屋类型，提供了一种更灵活的生活体验。

柔和住宅的结构采用了传统的实木结构和木钉结合的地板结构。实木结构外面还有一层太阳能纳米材料，它是一种轻质的智能获取能源的织物覆层，能改变自身角度来适应太阳照射角度或者使视野完全开放。织物覆层遮挡着可以看到一片风景区的巨大的玻璃幕墙。自然光可以通过一个三层楼高的空气对流中庭照射到一层，同时室内的遮帘和可控的通风口也可以调节温度的高低。

可移动的遮帘上的LED灯具有娱乐效果，让人们在人文和自然环境中间开展新的交流。柔和住宅智能遮帘的LED灯光效果会在固定的时间变换，并能随着外面的风向和气候状况的改变而改变。LED灯沿着柔和住宅的遮帘表面移动，会受外部风力的影响，形成一种视觉上的微风效果——用建筑自身华丽的灯光效果来表现外部的环境。

Soft House

The Soft House project in Hamburg, Germany is a winning competition entry for the International BauAustellung (IBA), a prestigious German building tradition that dates to 1901. Completed in March of 2013, it is a set of live/work row house units which offer a new model for low carbon construction and an ecologically responsive lifestyle that can be personalized to meet homeowners' needs. The Soft House demonstrates how domestic infrastructure can become "soft" – engaging flexible living concepts, carbon-neutral solid wood (brettstapel) construction, and wireless building controls with responsive and performative textiles which create the public identity of the architecture. Through the conceptual reframing of "soft" and "hard" materials and the integration of architecture, mobile textiles, and clean energy infrastructure, the Soft House transforms the German passive house typology, offering a more flexible living experience.

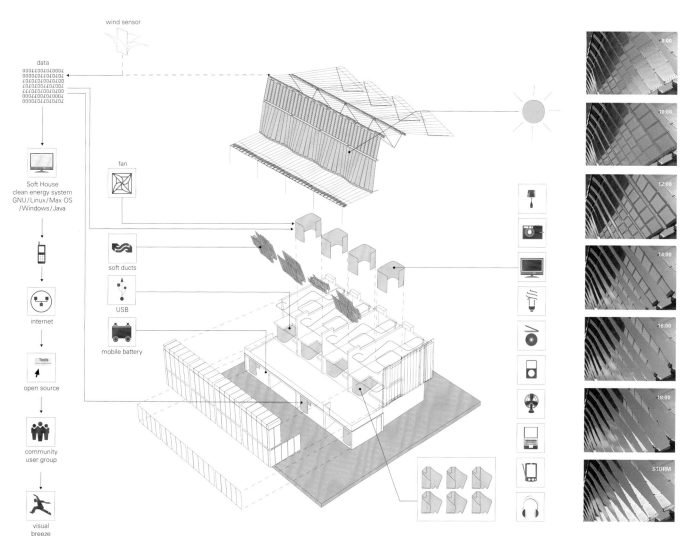

智能建筑管理系统 smart building management system

The Soft House's structure is a traditional solid wood panel and deck construction with wood dowel joints. The wood construction is complimented by flexible solar nano-materials in a lightweight, smart energy-harvesting textile cladding which bends to respond to sun angle or to open views. The textile cladding shades a large glass curtain wall with views to the parkland. A three-story air convection "atrium" brings daylight deep into the ground floor and modulates the rise and fall of warm and cool air with a system of interior curtains and operable window vents.

LED in the movable curtains are playful and engaging – allowing people to make new connections between the domestic and natural environments. The Soft House smart curtain LED lighting system allows for real time monitoring and visualization of outside wind and climate conditions. The solid state light that moves along the Soft House curtain surface in relation to exterior wind levels creates a Visual Breeze – an ambient interior luminous expression of the external environment.

Low-Carbon Flexible Photovoltaics
- Shared energy harvesting infrastructure provides 9,600 watts x 7 hours a day on average: over 24,000 kWh per year.
- Flexible Photovoltaics reduce manufacturing emissions by 40% when compared to glass PVs
- The larger solar aperture of flexible PV materials means that more energy is harvested in partly sunny or cloudy days.

Smart Shading Maximizes Energy and Views
- The twisters follow the sun through the day, providing shading and maximizing energy while preserving views.
- The translucent twister textile diffuses high-angle summer sunlight and allows shallow winter light to pass below.
- By following the sun's daily East-to-West path and making seasonal adjustments, the PV's efficiency increases 15%.

Carbon Neutral Construction
- The locally produced wood brettstapel structure reduces carbon emissions in transportation of building materials.
- The wood structure sequesters carbon that could enter the atmosphere, yielding a negative carbon footprint of approximately 180 metric tons of CO_2 (Hammond).
- A similar-sized concrete structure creates over 100metric tons of CO_2, a difference of 280 metric tons (Hammond).
- The flexible FRC board which supports the roof PVs are formed to naturally return to the "up" position. The recyclable aluminum structure and tension piston used to lower it require much less material than a compression structure.

Longevity and Adaptability of Architectural Structure
- The long lifecycle of the building reduces carbon output from repair/replacement construction.
- The flexible floor plan, with an open layout and multiple access points, can adapt to changes in program over time.
- The brettstapel structure contains no glue or nails, making it easily recycled at the end of its lifecycle.
- The solid wood structure is also a finish material. The resulting reduction in drywall eliminated over 6 metric tons of CO_2.

Transportation and Walkability
- The site is well served by public transportation within 150 ft, and a S-Bahn station within one-half mile.
- Each unit supports 4 bikes and one e-vehicle.
- The new IBA development has lifted the Walkscore of the Wilhelmsburg neighborhood to 80.
- The Soft House faces a traditional canal system which is used for recreation and transportaion.

Urban District-Level Clean Energy
- Wilhemsburg is powered by multi-source district-level renewable energy strategy initiated by the city. The Soft House block receives clean energy from a nearby landfill gas (methane) plant.
- Government initiated clean-energy development is now reduced due to the economic crisis. This makes on-site generation a more important component of carbon emission reduction.
- On-site power generation is flexible. Owners have the option to sell stored energy back to the grid at peak hours.

Innovative Low-Power, Clean-Energy Direct Current (DC) Ring
- The responsive PVs power a DC grid that provides energy to Smart Curtains and domestic appliances.
- By avoiding DC-to-AC converters, the grid design increases the total PV electrical output by 15%.
- The DC Ring is mounted on the wood ceiling rather than built into walls, eliminating material waste in infrastructure updates.

Extensive Natural Daylighting
- The south-facing atrium transforms the row house footprint, allowing daylight to penetrate deep into the ground floor.
- The open screen along the stairs encourages shared daylight between levels.

Energy Efficient Solid State Ambient Light
- The Smart Curtains and their integrated LED lighting can be moved to suit different activities, meaning fewer overall light fixtures are needed.
- Solid-state LED lighting can be twice as efficient as compact fluorescent light. This translates into a savings of 9 kg of CO_2 per 2000-lumen lightbulb per year.
- LEDs do not contain mercury and other toxic, non-recyclable materials that occur in fluorescent bulbs.

95 kWh/m²/yr - Predicted EUI, including on-site renewable energy
28 kWh/m²/yr - The solar contribution.
67 kWh/m²/yr - Predicted Energy Use Intensity excluding on-site renewable energy contribution.
182.3 kWh/m²/yr - Average EUI of a multifamily residence in the Northeastern United States

All Wood Construction (Brettstapel)
The Soft House design features a traditional all wood structure that utilizes sustainable soft wood spruce pieces pegged together without glue or nails. Fabricated by a local builder, the solid wood panels were shipped to the site and lifted into place quickly.

中间楼层遮帘主要位置
middle floor primary curtain movement position

unit A: position 1

unit A: position 2

unit B: position 1

unit B: position 2

unit B: position 3

unit B: position 4

unit C: position 1

unit C: position 2

unit C: position 3

unit C: position 4

unit D: position 1

unit D: position 2

unit D: position 3

unit D: position 4

上层遮帘主要位置
upper floor primary curtain movement position

unit A: position 1

unit A: position 2

unit A: position 3

unit A: position 4

unit B: position 1

unit B: position 2

unit B: position 3

unit B: position 4

unit C: position 1

unit C: position 2

unit C: position 3

unit C: position 4

unit D: position 1

unit D: position 2

unit D: position 3

unit D: position 4

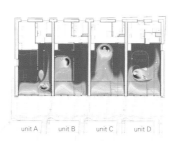

20° 19° 18°
辐射温度
radiant temperature

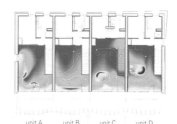

中间楼层室内热气候
middle floor interior thermal climate

上层室内热气候
upper floor interior thermal climate

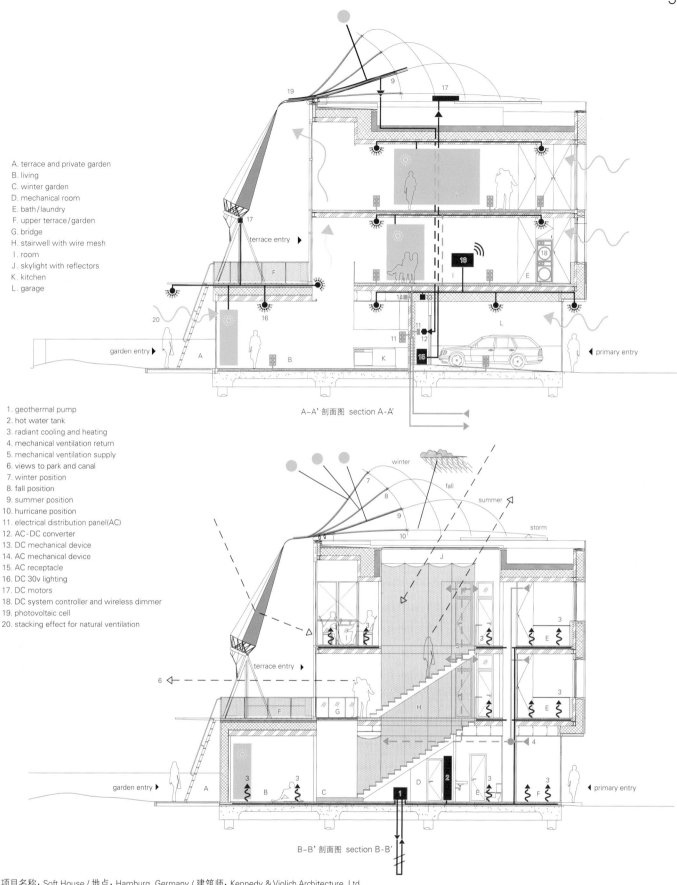

A. terrace and private garden
B. living
C. winter garden
D. mechanical room
E. bath/laundry
F. upper terrace/garden
G. bridge
H. stairwell with wire mesh
I. room
J. skylight with reflectors
K. kitchen
L. garage

1. geothermal pump
2. hot water tank
3. radiant cooling and heating
4. mechanical ventilation return
5. mechanical ventilation supply
6. views to park and canal
7. winter position
8. fall position
9. summer position
10. hurricane position
11. electrical distribution panel(AC)
12. AC-DC converter
13. DC mechanical device
14. AC mechanical device
15. AC receptacle
16. DC 30v lighting
17. DC motors
18. DC system controller and wireless dimmer
19. photovoltaic cell
20. stacking effect for natural ventilation

项目名称：Soft House / 地点：Hamburg, Germany / 建筑师：Kennedy & Violich Architecture, Ltd.
首席建筑师：Sheila Kennedy, Frano Violich / 高级经理：Veit Kugel / 设计团队：Kyle Altman, Jeremy Burke, Stephen Clipp, Iman Fayyad, Patricia Gruits, Katherine Heinrich, Heather Micka-Smith, Chris Popa, Shevy Rockcastle, Phillip Seaton, Alex Shelly, Nyima Smith, Sean Tang, Diana Tomova, Sasa Zivkovic
项目开发商：IBA _ IBA Hamburg GmbH, Soft House _ PATRIZIA Projektentwicklung GmbH / 现场监理建筑师：360grad+ architekten
景观建筑师：G2 Landschaft / 结构工程师：Knippers Helbig Advanced Engineering
气候规划：Steinbeis Forschungsinstitut für Solare, Zukunftsfähige Thermische Energiesysteme
暖通机械工程师：Buro Happold / 客户：Internationale Bauausstellung IBA Hamburg GmbH
用地面积：1,070m² / 建筑面积：370m² / 有效楼层面积：900m² / 设计时间：2012 / 施工时间：2013 / 竣工时间：2014
摄影师：©Michael Moser (courtesy of the architect)

巴西的生态商业建筑是德国在全球多国可持续发展建筑项目中的一部分，而作为一个休闲中心，它也展示了它的产品。创新的建筑材料、体系和技术是建筑设计的基础，并迎合了当地的生产条件，目标是在经济条件的约束下，尽可能地减少对环境的影响。

建筑师用半透明的聚碳酸酯板勾画出了矩形的轮廓，建筑内部的特点是容纳了不同的树木。泥土中的水分可以通过木质的平台和棚架渗透到地下。

建筑内有一个水池，能够收集雨水，有助于调节湿度和温度。水和电能的消耗通过建筑自动化系统控制，例如，该系统能够根据太阳光的照射角度来控制室内照明。

拜耳生态商业建筑

Loeb Capote Arquitetura e Urbanismo

Bayer-Eco Commercial Building

The Eco Commercial Building Brazil is part of the German multi-national's program of developing sustainable construction around the world and serves as a leisure center where it can also showcase its products. Innovative building materials, systems and technologies were the basis for the design project and adapted to the local conditions with the aim of producing as small an environmental impact as possible,

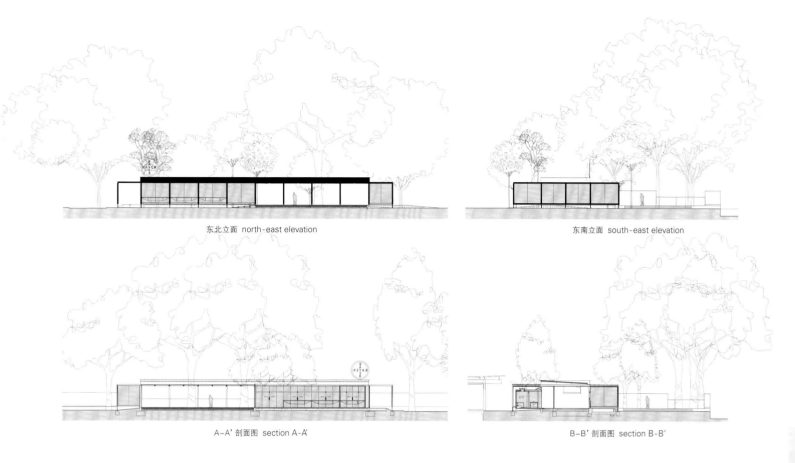

东北立面 north-east elevation

东南立面 south-east elevation

A-A' 剖面图 section A-A'

B-B' 剖面图 section B-B'

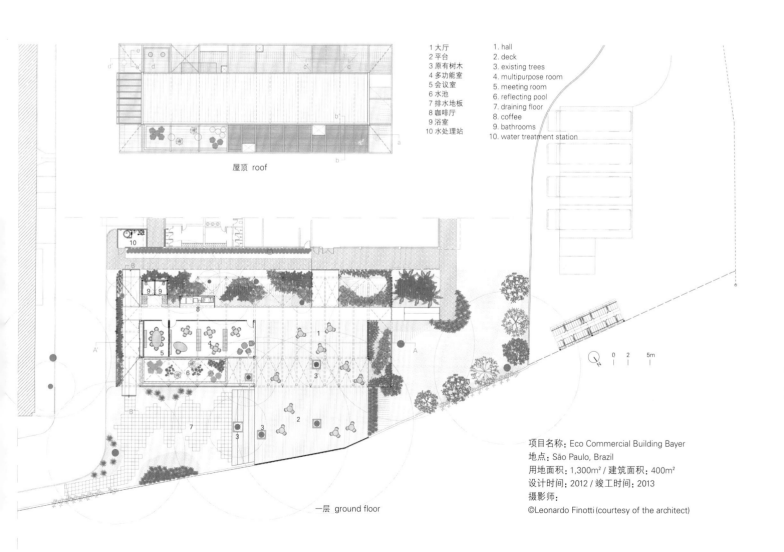

屋顶 roof

1 大厅　　　　1. hall
2 平台　　　　2. deck
3 原有树木　　3. existing trees
4 多功能室　　4. multipurpose room
5 会议室　　　5. meeting room
6 水池　　　　6. reflecting pool
7 排水地板　　7. draining floor
8 咖啡厅　　　8. coffee
9 浴室　　　　9. bathrooms
10 水处理站　 10. water treatment station

一层 ground floor

项目名称：Eco Commercial Building Bayer
地点：São Paulo, Brazil
用地面积：1,300m² / 建筑面积：400m²
设计时间：2012 / 竣工时间：2013
摄影师：
©Leonardo Finotti (courtesy of the architect)

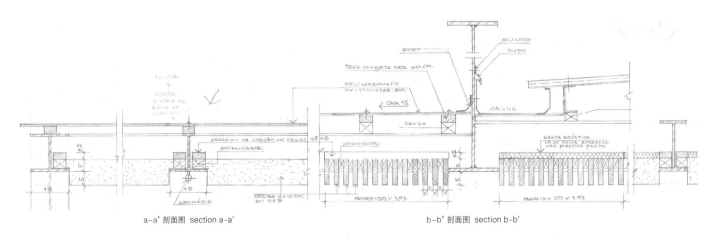

a-a' 剖面图 section a-a' b-b' 剖面图 section b-b'

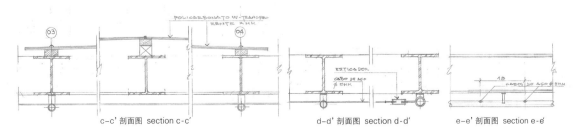

c-c' 剖面图 section c-c'   d-d' 剖面图 section d-d'   e-e' 剖面图 section e-e'

wooden pergola with low VOC content | ceiling in natural OSB wood urea-formaldehyde free | thermal isolation covering with polyurethane | closing in milky alveolar polycarbonate for natural light | photovoltaic panel

existing vegetation | metallic structure with low content of paint VOC | polyurethane floor with low VOC content | reflecting pool storage of treated rain water for reuse | wooden deck with resin treatment water base | closing with transparent polycarbonate

within the constraints of economic feasibility. The rectangular deployment is delineated by translucent polycarbonate planes, with the internal area featuring various trees. The soil retains its permeability through the presence of wooden decks and pergolas.

A reflecting pool collects rainwater and helps to regulate the humidity and temperature. Water and power consumption is monitored by a building automation system that controls the internal lighting based on the incidence of sunlight coming in, for example.

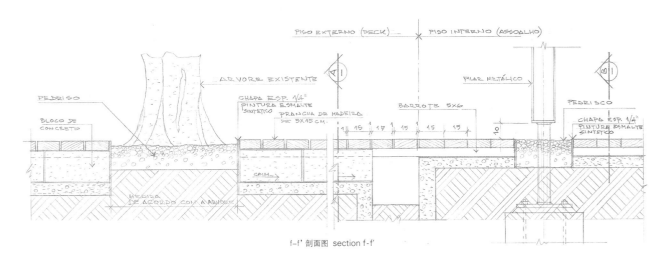

f-f' 剖面图 section f-f'

能源意识与可持续公共空间 Energy-conscious and Sustainable Public Spaces

亚利桑那州立大学卫生服务楼
Lake|Flato Architects

亚利桑那州立大学卫生服务楼是一个给学生提供温暖便利和帮助的基础设施建筑，它通过高效、功能齐全、完善学校自身组织结构的高效运转机制设计，营造了一种健康和幸福感并存的氛围。

项目包括拆除一部分原有设施，翻修一部分原有设施，并建造一座能改变整个服务楼的新建筑。

整座建筑将修建校园棕榈小径与修建高效并受欢迎的内部环境并重。穿过小型的侧厅，或经过前门风景优美的庭院和郁郁葱葱的门廊，都可以步入校园主要的步行街棕榈小径。新增建的区域与原有建筑协调搭配，采用相同的色调和建筑材料，既协调统一了整座建筑，又不失特色景观。

原有的门厅被拆掉了，建筑内部的公共通道被设在东边，与棕榈小径平行。公共通道的两侧修建了两层楼的侧厅，供病人休息等待。侧厅之间是一系列私密的景观庭院，这些区域作为室外等候区，可让身体健康的人到户外活动，与病人分隔开。

新的设计不仅赋予老旧破损的原建筑以活力和能源节约概念，还创新地增加了很多供学校老师和学生使用的便捷高效设施。服务楼吸收了新技术，包括增加了卫星保健诊所，不但能进行常规咨询，还能进行虚拟咨询，从而增加了接受病人病情咨询的能力，现在每天咨询的病人人数增加了7%。

服务楼中小规模的空间、天然的建材、充足的自然光以及公共场所和私密空间的有机结合，营造了一个健康的环境，增添了病人的幸福感，避免了人与人之间的冷漠相处。

这座建筑获得了绿色能源与环境设计（LEED）白金奖，医疗服务更是将综合医疗的措施（针灸、按摩、理疗和营养咨询）和传统的医治方式结合起来，确保病人能够得到最好的医治，让他们可以在亚利桑那州立大学痊愈。

Arizona State University Health Services Building

The Arizona State University(ASU) Health Services Building is a warm and inviting facility for students, creating a sense of health and wellness within an efficient, functional and timeless design that reinforces the campus fabric.

The project included the partial demolition of existing facilities, the renovation of the original facility and the construction of a new wing that transformed the identity of the facility.

Along with creating an efficient and welcoming interior, the building now engages ASU's historic Palm Walk, the campus' primary pedestrian spine. Integrated into a smaller-scale wing that steps out towards the Palm Walk, the new front door welcomes patients through a landscaped courtyard and shady porches. The remainder of the addition matches the scale of the original facility while incorporating a similar material palette that creates a unified facility and a clearly defined landscaped edge.

土地使用
land use

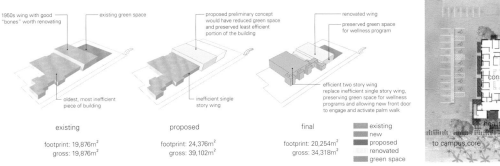
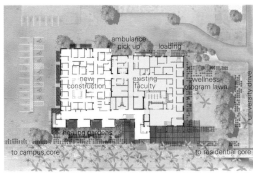

地区和社区特点
regional and community characteristics

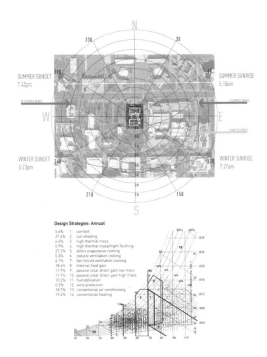

场地生态
site ecology

材料
materials

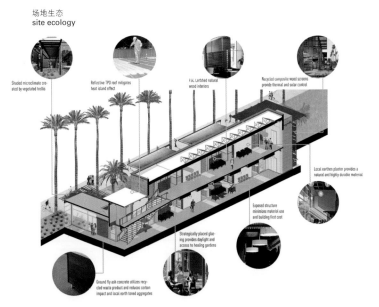
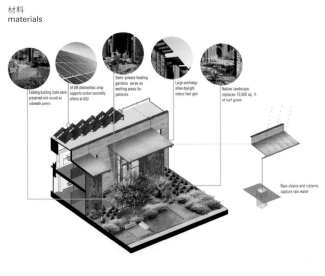

Working off the existing lobby, the interior public circulation is located on the east side paralleling with the Palm Walk. A series of two-story bays punctuate the linear public circulation providing intimate and private waiting areas for patients. These lower scaled bays frame a series of private landscaped courts that serve as exterior waiting areas, allowing healthy patients to be separated from sick patients and providing opportunity to engage with the outdoor spaces.

In addition to the transformation of the old dilapidated building into an engaging, energy-efficient facility, the new design has created more accessible and accommodating spaces for students and faculty that aid in workflow efficiency. The incorporation of new technology, including a satellite health clinic that allows for virtual consultations in addition to standard consultations, has increased the percentage of patient consultations per day by seven percent.

The intimate scale, natural materials, abundant natural daylighting and the combination of visibility and privacy foster a healthy environment that promotes wellness while avoiding a cold and clinical feel.

Achieving LEED Platinum certification, the building allows for the combined practice of integrative medicine (acupuncture, chiropractic, massage therapy and nutrition counseling) with traditional medicine to keep patients in the best of health, allowing them to be successful at ASU.

立面详图
elevation detail

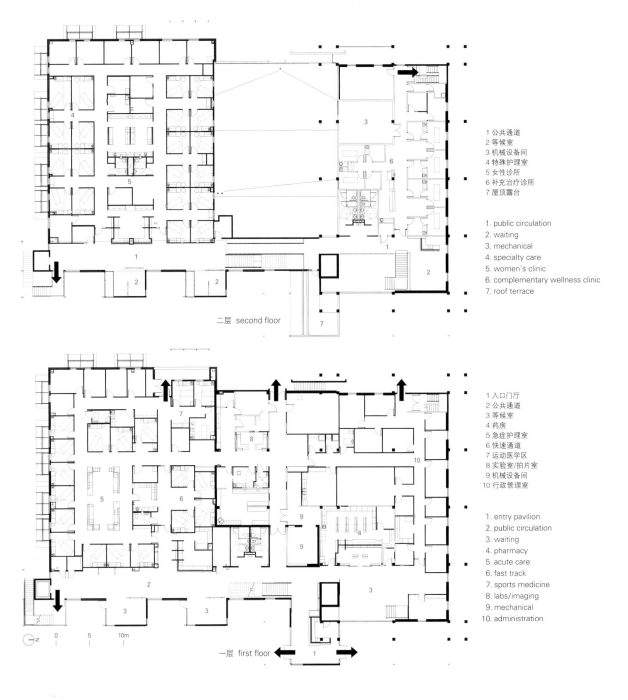

1 公共通道
2 等候室
3 机械设备间
4 特殊护理室
5 女性诊所
6 补充治疗诊所
7 屋顶露台

1. public circulation
2. waiting
3. mechanical
4. specialty care
5. women's clinic
6. complementary wellness clinic
7. roof terrace

二层 second floor

1 入口门厅
2 公共通道
3 等候室
4 药房
5 急症护理室
6 快速通道
7 运动医学区
8 实验室/拍片室
9 机械设备间
10 行政管理室

1. entry pavilion
2. public circulation
3. waiting
4. pharmacy
5. acute care
6. fast track
7. sports medicine
8. labs/imaging
9. mechanical
10. administration

一层 first floor

项目名称：Arizona State University Health Services Building
地点：451 East University Drive, Tempe, Arizona 85287, USA
建筑师：Lake|Flato Architects
提交方案建筑师：Orcutt|Winslow
室内设计师：Orcutt|Winslow
机电工程师：Van Boreum & Frank Associates
结构工程师：Caruso Turley Scott Inc.
土木工程师：Littlejohn Engineering Associates
施工经理：Okland Construction
总承包商：Okland Construction
景观建筑师：Ten Eyck Landscape Architects
绿色设计咨询/LEED咨询/生命周期性能合伙人：Orcutt|Winslow
有效楼层面积：34,318m²
竣工时间：2012.5
摄影师：©Bill Timmerman (courtesy of the architect)

大卫与露西尔·帕克基金会总部

EHDD

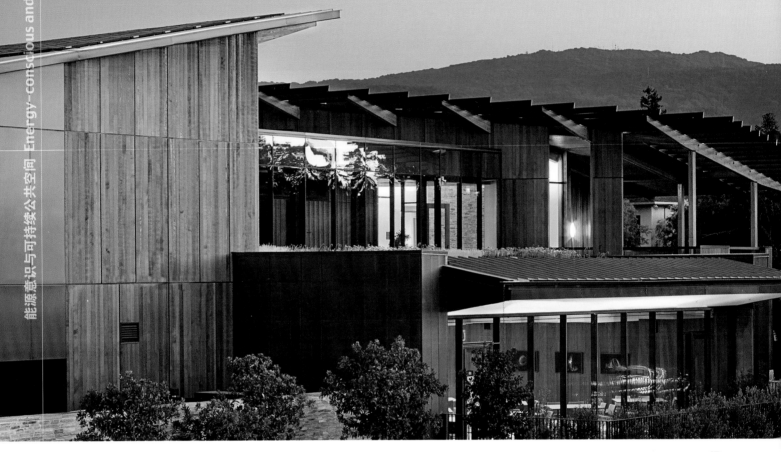

大卫与露西尔·帕克基金会总部是助推全世界性可持续发展能力，带领其员工、受益者和合伙人携手解决世界上最棘手的可持续发展问题的组织。基金会和Los Altos社区的联系可以追溯到1964年基金会刚成立时。在过去的20年里，基金会的工作内容从地区扩展到全世界，员工和办公地点分散在城市中的多座建筑中。这个项目增加了基金会员工之间的互动和合作机会，对当地市中心一个项目建设的投资预计会持续到21世纪末，从而增加了与当地民众的接触。

这座建筑根据加利福尼亚州温和而独特的气候以及自然光照而设计，还有多条通向室外的通道。两个12m宽的工作室模块分开设置，围出了一个有当地自然风景特色的室外中庭，这个中庭是整个设计中最大的房间，一年中大部分的时间被作为会议室使用。建筑功能空间围绕中庭设置以保证最大的通透度，两座玻璃桥分隔两个楼体（两个楼体内容纳的是合作空间），从外观上保证了最大程度的透明度，同时打破了两个楼体从街道上看上去的规模。所有的用户都能通过可开启窗获得阳光，中庭几乎成为所有用户的中心元素，无论用户身处建筑中何处，都能感觉到它的存在。中庭在一层的四周设有推拉隔断，这些隔断模糊了室内外的界线，开放程度取决于温度、昼长和工作人员的需求。行之有效的工作进程会产生计划统一的工作环境。12个模块式的工作室可以灵活地根据不同工作组、不同工作时间的需求，进行统筹安排。11m²的标准办公室/会议室和7.4m²的写字间相邻，提供了能满足不同工作功能需要的空间。

客户希望拥有一座低调的建筑，外观介于中等办公楼和大型家庭住宅之间，与粗糙的工业建筑相比，客户更喜欢有温暖色调的建材，建筑用细致有序的方式表现了树木、石头、铜和玻璃的自然之美，极致展现了自然的本质和这些材料的永恒。这座建筑完美地融入了Los Altos的周围环境，它是一座朴实无华的当代建筑，与Los Altos和其他硅谷社区中的非住宅建筑相比并不那么抢眼。

设计中的运输需求管理计划，减少了800万美元的地下停车位建设费用，同时避免了限制单车通勤。建筑设计理念基于可复制性：建筑的最终目标不是最低成本，而是用十年的时间证明它是市场上具有价格优势的最好的建筑。这座建筑现在已经产生了比最初两年总消耗达多的能源。最终，这个设计证明，一座更具有可持续发展性的建筑能够丰富人们的日常生活，让人们可以更好地配合来解决世界上最棘手的问题。

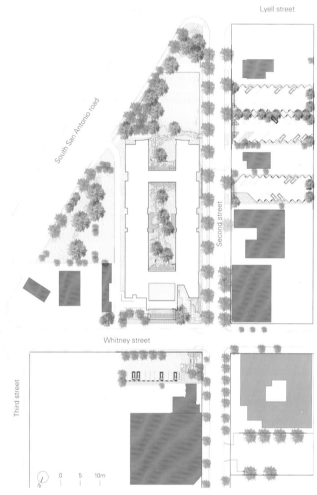

The David and Lucile Packard Foundation Headquarters

The David and Lucile Packard Foundation Headquarters acts as a catalyst for broad organizational sustainability and brings staff, grantees and partners together to solve the world's most intractable problems. The Foundation's connection to the Los Altos community dates back to its inception in 1964. For the last two decades, as its grant making programs expanded locally and worldwide, staff and operations have been scattered in buildings throughout the city. This project enhances proximity and collaboration while renewing the Foundation's commitment to the local community by investing in a downtown project intended to last through the end of the 21st century.

The building is tuned for a uniquely benign California climate and shaped by daylight and access to the outdoors. Two 12m wide workplace modules were pulled apart to create an indigenously landscaped outdoor atrium which functions as the project's largest room and active meeting space most of the year. The program was organized around the atrium to allow for maximum transparency through the site – by way of two glass bridges separating the wings that house collaboration spaces – in order to break down the scale of the building when viewed from the street. All occupants are within easy reach of daylight and operable windows and the atrium exerts an almost centripetal force on the occupants – its presence felt at every turn. The demarcation between indoors and outdoors is blurred with the help of sliding partitions at the atrium's ground level perimeter, the degree of openness

北立面 north elevation

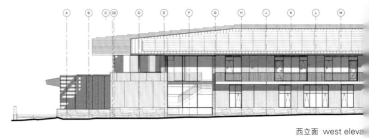
西立面 west elevation

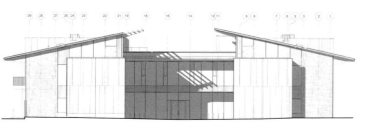

南立面 south elevation

1 接待室	1. welcome room
2 门厅	2. lobby
3 大厅	3. foyer
4 大会议室	4. large meeting room
5 办公室	5. office neighborhood
6 开放式会议室	6. open meeting
7 会议室	7. meeting
8 员工休息室	8. staff lounge
9 厨房	9. kitchen
10 露台	10. terrace
11 董事会会议室	11. board room
12 复印室	12. copy
13 建筑设备间	13. servers
14 机械设备区	14. mechanical area
15 绿色屋顶	15. green roof

二层 second floor

一层 first floor

项目名称：The David and Lucile Packard Foundation
地点：343 Second Street, Los Altos California 94022, United States
建筑师：EHDD / 首席建筑师：Scott Shell
首席设计师：Marc L'Italien / 项目经理：Brad Jacobson
项目建筑师：Terry McCormick _ exterior, Lotte Kaefer _ interior
总承包商：DPR CONSTRUCTION, Inc. / 室内设计：Michelle Hill _ EHDD
景观设计：Joni L. Janecki & Associates
土木工程师：Sherwood Design Engineers
结构工程师：Tipping Mar / 机械/管道/电气工程师：Integral Group
规范咨询：The Fire Consultants / 照明设计：Janet Nolan & Associates
声学设计：Charles M. Salter & Associates, Inc.
控制系统(SCADA)：Pipeline Systems (PSI)
客户：The David and Lucile Packard Foundation
建筑面积：49,161m² / 竣工时间：2012.7
摄影师：©Jeremy Bittermann (courtesy of the architect)

depending on temperature, time of a day and staff's needs. An engaged programming process resulted in a plan supporting collaboration in the workplace. Twelve modular office neighborhoods provide flexibility for working groups as they change over time. Standardized, 11m² flex office/meeting rooms and 7.4m². ft. workstations are the neighborhoods' building blocks, providing a range of spaces suitable to diverse work functions.

The client desires a building that expresses these humble roots with an appearance that lies somewhere between a modest office building and a large family home. With a disdain for a techy, industrial aesthetic, the client encourages a warmer palette of materials. The architecture expresses the natural beauty of wood, stone, copper and glass, detailed and assembled in a taut, minimal manner that best showcases the subtle nature of naturally-finished, time-honored materials. The result is a project that feels at home in Los Altos and advocates an honest, contemporary, and less derivative aesthetic for non-residential architecture of this and other Silicon Valley communities.

A transportation demand management plan helped eliminate the need for an $8 million underground parking garage and dis-incentivized single-car commuting. Design choices were passed through a "replicability" filter: the goal was not the least cost, but to identify best practices that within a decade were poised to transform the building marketplace at a competitive cost. The building we created has produced more energy than it has consumed in its first two years of operation. Ultimately, the project demonstrates that a more sustainable life is simply a better one by enriching people's daily life and their ability to work together to solve the world's most intractable problems.

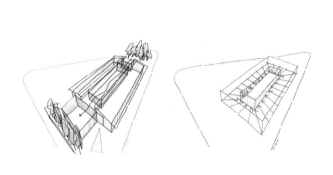
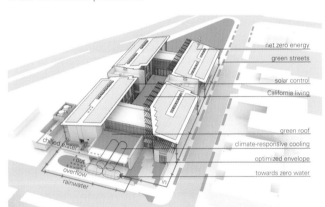

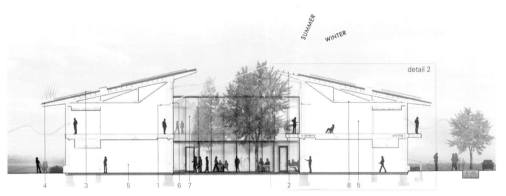

1. light shelves reflects winter sun, maximizing daylighting 2. trellis filters hot, summer sun providing courtyard shading
3. PV array results in net positive energy balance 4. captured rainwater or toilet flushing and irrigation
5. 12m wide maximized daylighting and natural ventilation 6. interior/exterior blinds control sunlight
7. triple-glazed, highly insulating windows reduce heating system 8. exposed FSC certified wood structure

A-A' 剖面图 section A-A'

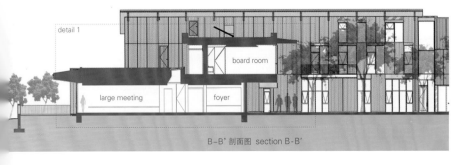

B-B' 剖面图 section B-B'

C-C' 剖面图 section C-C'

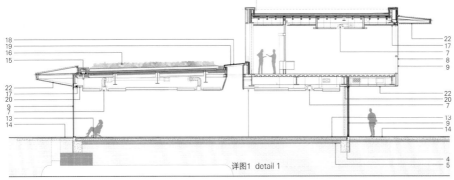

详图1 detail 1

building systems

1. photovoltaic panels supply 100% of energy
2. R19 wall insulation
3. R31 roof insulation
4. slab edge insulation
5. under-slab insulation
6. light shelf with integrated radiant cooling panels
7. active chilled beams provide heating and cooling
8. manual operable windows-sensors notify users when to open/close windows
9. triple glazed, highly insulating windows
10. energy efficient lighting on daylight sensors

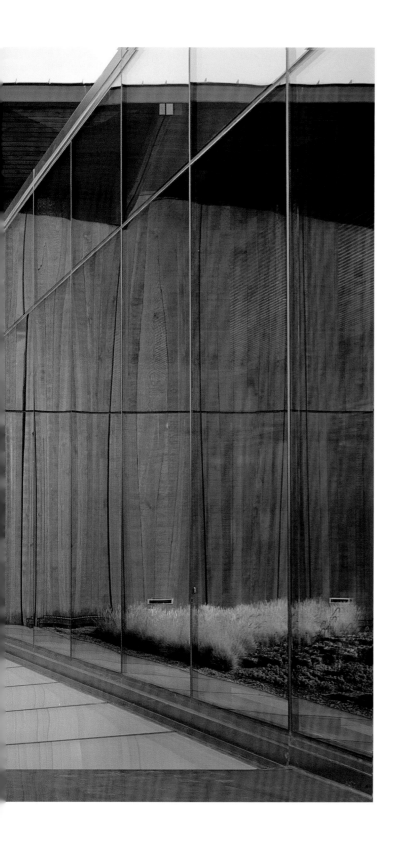

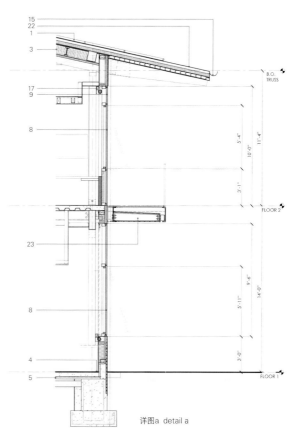

详图a detail a

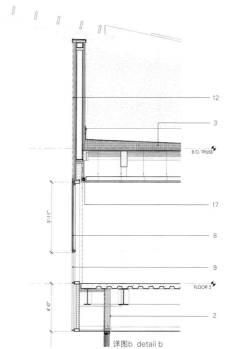

详图b detail b

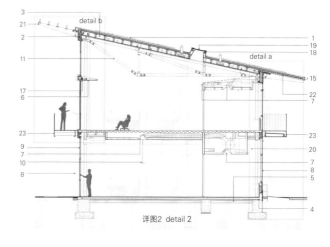

详图2 detail 2

materials
11. exposed timber structure
12. copper siding
13. natural stone flooring
14. natural stone paving

rainwater/landscaping
15. rainwater collected for reuse
16. water saving landscaping
17. interior automated shades provide sun control
18. skylights provide natural daylighting
19. fritted glass provide sun control at skylights
20. maximized daylighting solar-tubes provide natural daylighting
21. trellis allows winter sun light into courtyard
22. wide eaves provide sun control
23. balconies provide sun control

ECCO酒店与会议中心

Dissing+Weitling Architecture

ECCO酒店与会议中心的设计目标是给人们提供一种建筑体验,体现ECCO品牌的价值与传统。建筑的设计在各个方面都几近完美,不只是地热能和太阳能的使用,还有提高空间使用效率的环形楼层平面设计,既缩短了建筑内部的距离,又减少了建筑表面积,将热量的损失减至最小。而且建筑的主要材料就是水泥,石膏板和木头。酒店客房内天花板上特别设计的嵌入式LED射灯,让人感觉光线是从混凝土内发射出来的。

所有的公共区域都设计了斯德哥尔摩灯具,这些灯具被无缝融合在隔声石膏板天花板上。圆形的建筑形式而非复杂的几何结构更有利于各种固定设施的安装。酒店建筑就是一座能源盈余建筑,它不仅仅可以自给自足,而且在一年中的特定时间,还能生产剩余能源,供给新建筑附近的原有会议中心使用。

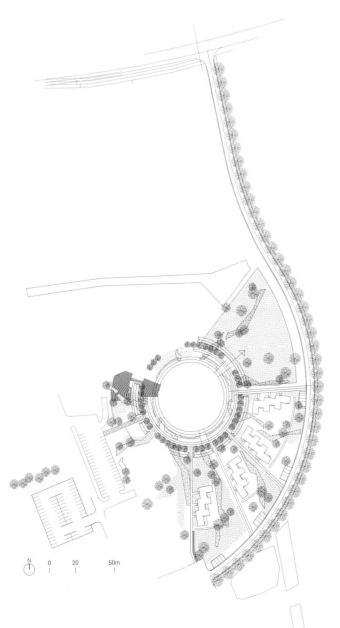

东立面 east elevation

北立面 north elevation

项目名称：ECCO Conference Center / 地点：Ecco Alleen 4, Tønder, Denmark / 建筑师：Dissing+Weitling architecture
工程师：INGENIØR'NE / 景观建筑师：Yards Landscape / 室内设计：FORM Cph / 客户：ECCO / 用地面积：4,000㎡ / 有效楼层面积：5,000㎡
设计时间：2012 / 施工时间：2012 / 竣工时间：2013 / 摄影师：©Adam Mørk (courtesy of the architect)

酒店建筑中容纳了多种功能的房间，在同一座建筑中能同时实现客房、展览室、会议室、多媒体房间和休闲娱乐区，在此类型建筑中是非常少见而独特的计。这样紧凑的设计也让我们有机会去思考建筑中的不同分区（私密、半私密和公共区域）应该如何分布，才能让用户有更多的或正式或非正式交流的机会。

将私密的独立客房分布在建筑外围的设计，能够保证每一个房间都能有充足的阳光，同时可以欣赏到周围的风景，还有一个小小的室外私密空间。客房既是一个私密的休息空间，又是一个功能性的工作空间，还可以是一个私人的社交场所。

建筑整体的重心，或者说建筑的中心，是一个两层楼高的多功能大厅。会议、时装展、授课或晚宴都可以在这个明亮、宽敞并且多功能的空间内举行。无论大厅内一个人没有，还是因为有活动而装满了人，其空间的质量、选用的建材、自然光的应用和仔细设计的声学效果，都能让身处其中的客人身心舒畅。多功能大厅连接着很多展览室和私人会议室，呈现出了无缝的视觉连接和空间的功能共享。

环形的楼层平面被分为四个部分，灵感来源于世界不同的四种文化。根据内部装潢和总体感觉，四个部分的主题分别是斯堪的纳维亚、亚洲、非洲和美洲。在室内，这些主题主要表现在家具和装饰品的设计、颜色和选材上；在室外，则主要体现在整体色调和植物的来源上。

建筑材料的选择体现了ECCO使用自然资源的理念。地面采用橡胶和板条拼花地板，墙面则是木材和混凝土，酒店客房中的照明灯包裹在ECCO自己员工亲手缝制的皮革中。Dissing + Weitling建筑事务所转变了建材、工艺和设计的质感，将这些渗透到建筑设计中，既加强了ECCO的品牌效益，又能使用户直观体验到公司的价值。

ECCO Hotel and Conference Center

The design of the new hotel and conference center is based on a strong desire to create an architectural experience that exemplifies the values and traditions of the ECCO brand. The design of the building is optimized in every way, not only through the use of geothermal heating/cooling and solar energy but also the circular shape of the floor plan itself which makes for the best utilization of the available space, short distances within the building and minimum heat loss on account of the reduced surface area. Each solution works with the material it is mounted into – concrete, gypsum plaster, or wood. Special LED spots are cast into the concrete ceilings of the hotel rooms giving the impression that the light comes from the concrete.

For all the public areas, Stockholm lighting has developed that can be seamlessly integrated into the acoustical plaster ceilings. The round form allows the fixtures to fit easily into the buildings otherwise complex geometry. The hotel building can be classified as a plus energy building in as much as it is not only self-sufficient, but at certain times of the year it will actually produce a surplus of energy which will allow a small contribution to the consumption of the original conference center adjacent to the new building.

The combination of diverse functions, hotel rooms, showrooms, meeting rooms, multipurpose room, and lounge areas in one compact entity is very unusual and quite unique for a building of this type. The compact solution provided opportunities to rethink how the various zones of the building

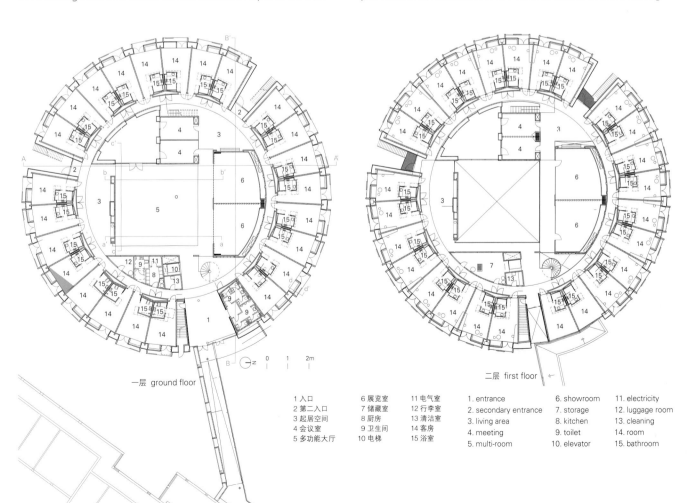

一层 ground floor
二层 first floor

1 入口	6 展览室	11 电气室	1. entrance	6. showroom	11. electricity
2 第二入口	7 储藏室	12 行李室	2. secondary entrance	7. storage	12. luggage room
3 起居空间	8 厨房	13 清洁室	3. living area	8. kitchen	13. cleaning
4 会议室	9 卫生间	14 客房	4. meeting	9. toilet	14. room
5 多功能大厅	10 电梯	15 浴室	5. multi-room	10. elevator	15. bathroom

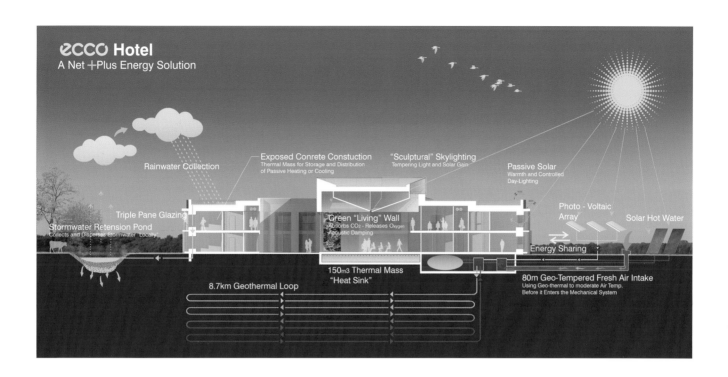

1. fins over exterior stairs
2. window above multi-room
3. ceiling above multi-room
4. glass towards green patio
5. skylight above showroom
6. skylight above green wall
7. skylight above water art
8. concrete wall with texture
9. glass towards green patio
10. concrete element wall

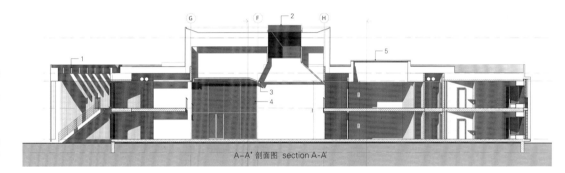

A-A' 剖面图 section A-A'

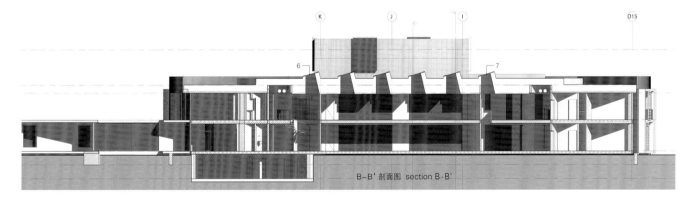

B-B' 剖面图 section B-B'

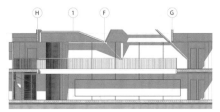

a-a' 剖面图 section a-a'

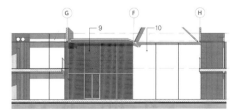

b-b' 剖面图 section b-b'

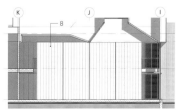

c-c' 剖面图 section c-c'

(private, semi-private and public) could complement each other and thereby facilitate a variety of opportunities for both formal and informal interaction.

By placing the individual (private) rooms around the periphery of the building, each room has the maximum amount of daylight, panoramic views to the surrounding landscape, and the opportunity for a small outdoor private space. The room is a private retreat, a functional work space, and a place for socializing at an individual level.

The center of gravity, or the heart of the building, is the two-floor-high multipurpose room. Conferences, fashion shows, lectures and dinners are all possible in this light, airy and super functional space. The quality of the room, its materials, natural daylight and carefully considered acoustics make it equally as interesting and comfortable when the room is empty as it is when it is full of activity. Showrooms and more intimate meeting rooms are placed in direct connection to the multipurpose room allowing for a seamless visual and functional interaction between the spaces.

The circular floor plan is divided into four sections, each of which takes inspiration from one of the four corners of the world. In terms of interior decor and general feel, the four sections are themed according to Scandinavia, Asia, Africa and America respectively. Indoors, the themes are expressed discreetly in small variations in furniture and furnishing, colors and materials, and outdoors, in the colors and origins of the plants and bushes.

The choice of materials reflects ECCO's philosophy of using natural products. The flooring is rubber and stave parquet, the walls are made from wood and concrete, and the light fittings in the hotel rooms are covered in leather sewn by ECCO's own employees. Dissing+Weitling Architecture has turned the quality of the materials, craftsmanship and design into the pervading themes of the building, strengthening the ECCO brand and providing the users with a sensuous experience of the values of the company.

1. facade cladding, moelven canadian ceder wood, brushed surface, vertical, BXD=CA.25x70mm, not fire treated/classe 1 cladding
2. tombak cladding, pre painted, hidden montage
3. wooden windows/doors
4. FER-not wooden cladding, moelven ceder wood, 13x135mm, smooth, oiled fire proof / classe 1 cladding
5. movable wooden fins
6. fins: moelven ceder wood, 21x45mm, rhombe shaped, guide for wooden fins oart steel, aligned in ceiling, floor
7. glas-alu facade closure, AS FA. schuco system FW50+SI, VAR2 anodized in dark special color, telescope solution in top by all slab connections lodpost profiles over several floors interrupted (hidden) at all slabs-demands for sound isolation
8. rail, frame in plain steel, handrail in ceder wood, sceptre in round steel, all steel varm galvanized
9. 2x12.5mm plaster, visible surface painted, on steel rigelsystem as gyproc, moisture barrier
10. mineral wool, subconstruction not shown
11. concrete construction/concrete slab/concrete wall, as ingeneer project
12. steel construction as ingeneer project
13. thermal glass, 3 layers, U-value=0.5, safety glass as DS/EN12600 /DS/NF119
14. MDF-plate, painted
15. cushion, leather padded
16. convector
17. leca-foundation block
18. cantilevered wood lock construction, underlays, classe 1 material, fixed to underlying wood lock construction permeable, fire treated
19. underlays, classe 1 material, fixed to underlying wood lock construction permeable, fire treated
20. shuttering: perforated hatprofiles or cross shuttering for ventilation of cladding
21. mural covering, zinc, pre-patinated, connected modular with sub plate and brackets

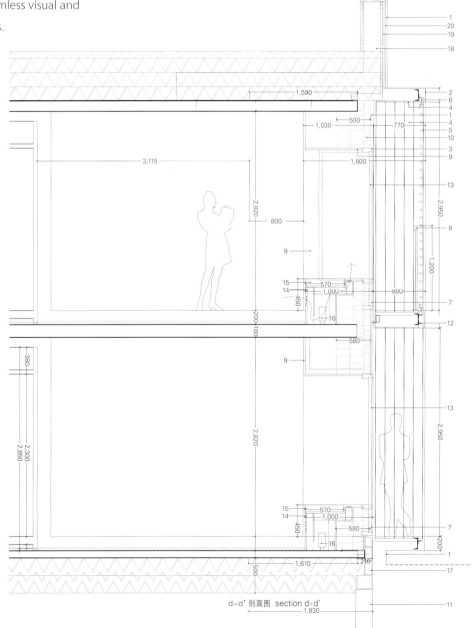

d-d' 剖面图 section d-d'

1 教室
2 实验室
3 行政管理办公室
4 宿舍
5 运动场
6 剧院花园
7 莲花池
8 诗意花园
9 竹园
10 自行车停车场

1. classrooms
2. laboratories
3. administration
4. dormitory
5. sports field
6. theater garden
7. lotus pond
8. poetry garden
9. bamboo garden
10. bicycle parking

北京第四中学房山校区

OPEN Architecture

著名的北京第四中学成立了分校区，位于北京西南五环外的新城中心区，占地4.5ha。新城区的总体规划理念是健康和自给自足，从而避免初期单一功能的城郊发展问题，而这座新校区就是城郊城市化进程的重要推动力。

校舍营造了一种更开放、更具有自然气息的环境氛围，这些都是当代中国城市学生最需要的。同时由于空间的限制，学校主要采用多层建筑的设计，并在建筑的上下层之间插入花园。这样的上下层建筑平行分布，并用多种方式连通到中心花园的独特的设计，象征的就是新学校中止式和非止式教育空间之间的关系。

下层建筑是大规模的非重复学校公共设施，包括食堂、礼堂、体育馆和游泳馆。其中的每一个区域都有各自不同的高度要求，从低到高排列建造，一直向上延展到可以接触到上层建筑的底部；因为这样的高度变化，它们的屋顶就形成了一个波浪式的室外花园。上层建筑的轮廓就像一棵根茎植物，主要是教室、实验室、宿舍和行政管理部门等重复空间。它的叶子都是向外延伸的、弯曲的或者突出的，但都通过根茎相互连接。连接这些房间的就是宽阔的主走廊，学生在课间的时候可以快速通过走廊。上层也有一些小组活动专用的半封闭的空间，这些就像是一条大河中有机形状的小岛。上层建筑的屋顶被设计成了有机农场，有36块土地分给学校的36个班级，提供给学生学习耕种技巧、体验田园生活的机会。

上下两种不同教育空间的对比，造就了这个令人叹为观止的复杂的空间分布。每一个空间都有自己的特征，从而使这个复杂的教育设施给人一种身处城市中的体验。不像有明确的空间分布，通常是有中轴线的对称分布的传统学校，这所新学校有自己自由的排列形式，有多个中心，可以从不同的方向进入学校。它是一个拥有自由灵魂的地方，鼓励每一个学生的探索和再创造。这样充满希望、充满生机的环境，可以慢慢改变中国的教育体系。

这个项目的设计目标是成为第一个位于城郊的三星级绿化学校（超过LEED金奖）。校舍最大限度地利用了自然通风和自然采光，并在建筑几何构造规划直至窗户细节设计等方面，都采用了被动式太阳能设计，减少夏季的热获得和冬季的热损失，保证内部达到冬暖夏凉的效果。可渗透的铺地材料和拓展型绿化屋顶都将表面径流降到最低。三个大的地下蓄水池从运动场搜集珍贵的雨水来灌溉有机农场和花园。转化地热能的热泵持续供给大型公共空间所需的电力，教学空间则可以自行操控中央空调系统来灵活满足不同的需求。整个项目都采用简洁、自然和持久的材料：竹胶合板、碎石（一种消失的工艺）、石头和裸露混凝土。当代中国讨论最多的问题就是个人、社会和自然的关系，教育对此承担了巨大的责任。创新的校舍既是对这些挑战的回应，也是一种尝试。

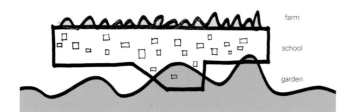

Beijing No.4 High School Fangshan Campus

Situated in the center of a new town just outside Beijing's southwest fifth ring road, this new public school on 4.5 hectares of land is designed as the branch campus for the renowned Beijing No.4 High School. As an important piece in a grand scheme to build a healthier and self-sustainable new town, avoiding problems of the earlier mono-functional suburban developments, the school is vital to the newly urbanized surrounding area.

The intention of creating more open spaces filled with nature, something that urban Chinese students today desperately need, combined with the space limitations of the site, inspired a strategy on the vertical dimension to create multiple grounds, by separating the programs into above and below, and inserting gardens in-between. The juxtaposition of the resultant upper and lower building, connected at the "middle-ground" in various ways, is as much an interesting spatial strategy as a signifier of the relationship between formal and informal educational spaces in the new school.

The lower building contains large and non-repetitive public functions of the school, such as the canteen, the auditorium, the gymnasium, and the swimming pool. Each of these spaces with their varying height requirements, pushes the ground up from below into various mound shapes that touch the belly of the upper building; their roofs in the form of landscaped gardens become the undulating new open "ground". The upper building is a thin rhizome shaped slab that contains the more repetitive and rigid programs of classrooms, labs, dormitories and administration. Its mega form extends, bends, and branches, but all connected together. The main circulation spine within this mega structure is widened to

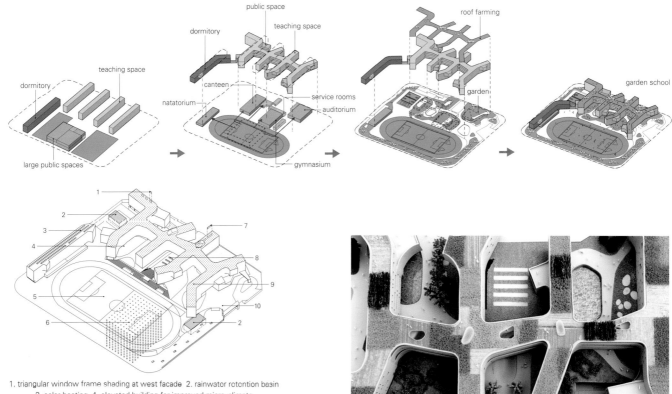

1. triangular window frame shading at west facade 2. rainwater retention basin
3. solar heating 4. elevated building for improved micro-climate
5. sunken sports field 6. geothermal ground source heat pump
7. sunshade window frame at south-facing classrooms 8. vented skylights incorporated into benches 9. planted roof 10. permeable paving

allow rapid foot traffic during class breaks. It also accommodates some semi-enclosed spaces for small group activities, like a river with organic shaped islands. The roof-top of the upper building is designed to be an organic farm, with 36 plots for the 36 classes in the school, providing students the chance to learn the techniques of farming, and also paying tribute to the site's pastoral past.

The contrast between the two types of educational spaces and the rich mix of programs within create a surprising spatial complexity. With the unique character for each different space, an urban experience is created within this complex of education facilities. Unlike a typical campus with hierarchical spatial organization and often clear axis to organize more or less symmetrical movements, this new school is of free form and meant to have multiple centers that can be accessed in any possible sequences. It is a place with a free spirit that encourages explorations and awaits reinventions by different individuals. Hopefully the physical environment can inspire and initiate some much needed changes in the education system of China today.

This project aims to be the first triple-green-star rated school in the country (a standard that exceeds LEED Gold). In order to maximize natural ventilation and natural light, and minimize heat gain during summer and heat loss in the winter, passive solar strategies are adopted in almost all aspects of the design, from the planning of the building geometry all the way to the details of the window design. Permeable ground surface paving and expansive green roofs helps to minimize surface run-off, and three large underground water retention basins collect precious rain water from the athletics field for irrigation of the farms and gardens. A geothermal ground-source heat pump provides a sustainable source of energy for the large public spaces, whilst independently controlled VRV units serve all the individual teaching spaces to ensure flexible operation. Throughout the project, simple, natural, and durable materials such as bamboo plywood, pebble dashing (a vanishing technique), stone, and exposed concrete are used. In the contemporary Chinese context, arguably the most pertinent issue and challenge is that of the relationship among individual, society, and nature. Education bears great responsibilities. It is to these issues that this new campus project aspires to be both a touchstone and a response.

二层 second floor

1 教室	1. classroom
2 机械设备间	2. mechanical room
3 展览空间	3. exhibition space
4 演讲厅	4. lecture hall
5 卫生间	5. restroom
6 小岛	6. island
7 活动空间	7. activity space
8 教师办公室	8. teachers' office
9 喷水池	9. water fountain
10 会议室	10. meeting room
11 储藏室	11. storage room
12 行政管理办公室	12. admin office
13 实验室	13. laboratory
14 准备室	14. preparation room
15 休息区	15. lounge area
16 公共阳台	16. public balcony
17 公共空间	17. public space
18 公共卫生间	18. public washroom
19 宿舍	19. dormitory
20 公共淋浴间	20. public shower
21 洗衣房	21. laundry room

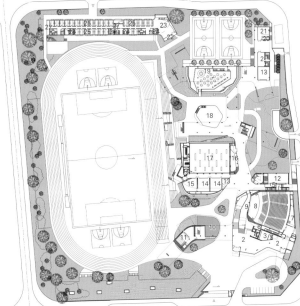

夹层 mezzanine

1 门卫室	1. guard's room
2 大厅	2. lobby
3 VIP室	3. VIP room
4 卫生间	4. restroom
5 储藏间	5. storage room
6 便利店	6. convenience store
7 投影室	7. projection room
8 礼堂	8. auditorium
9 活动空间	9. activity space
10 池塘	10. pond
11 演讲厅	11. lecture hall
12 舞蹈室	12. dance room
13 机械设备间	13. mechanical room
14 音乐教室	14. music classroom
15 教师办公室	15. teachers' office
16 攀岩室	16. rock climbing
17 健身房	17. gym
18 竹园上方空间	18. above the bamboo garden
19 教工食堂	19. faculty canteen
20 值班室	20. duty room
21 教师休息室	21. teachers' lounge
22 公共阳台	22. public balcony
23 公共空间	23. public space
24 公共卫生间	24. public washroom
25 宿舍	25. dormitory
26 公共淋浴间	26. public shower
27 洗衣房	27. laundry room

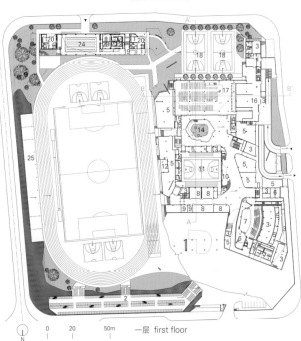

一层 first floor

1 门卫室	1. guard's room
2 自行车停车场	2. bicycle parking
3 机械设备间	3. mechanical room
4 卫生间	4. restroom
5 储藏间	5. storage room
6 后台	6. backstage
7 礼堂	7. auditorium
8 工作室	8. workshop
9 隔离病房	9. quarantine room
10 攀岩室	10. rock climbing
11 健身房	11. gymnasium
12 广播室	12. broadcasting room
13 身体检查室	13. physical examination room
14 竹园	14. bamboo garden
15 便利店	15. convenience store
16 厨房	16. kitchen
17 学生食堂	17. student canteen
18 篮球场	18. basketball court
19 游泳馆大厅	19. swimming pool lobby
20 宿舍大厅	20. dormitory lobby
21 公共卫生间	21. public washroom
22 更衣室	22. changing room
23 淋浴间	23. shower
24 游泳馆	24. swimming pool
25 运动设施房间	25. sports facility room

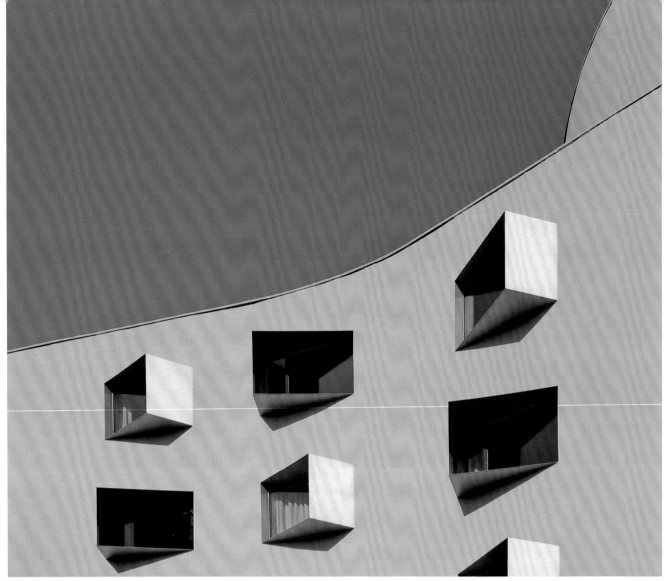

项目名称：Garden School / Beijing No.4 High School Fangshan Campus / 地点：Changyang, Fangshan District, Beijing
建筑师：Open Architecture / 首席建筑师：Li Hu, Huang Wenjing / 项目团队：Daijiro Nakayama _ 项目建筑师, Ye Qing, Zhang Hao, Zhou Tingting, Thomas Batzenschlager, Zhang Chang, Jotte Seghers, Wang Yifan, Li Qiang, Ge Ruishi, Xue Wencan, Brendan Whitsitt, Ami Kito, Tao Wei, Simina Cuc, Zhao Yao, Chen Xiaoting, Cynthia Yurou Cui, Tang Wei; Yu Qingbo, Felipe Escudero, Julia Mok, Lu Chen, Scott Craven
当地设计机构、结构工程师、机械工程师：Beijing Institute of Architectural Design / 施工管理：Vanke COFCO
可持续发展顾问：School of Architecture, Tsinghua University / 幕墙顾问：Inhabit Group
照明顾问：Lighting Design Partnership International / 声学顾问：Clocell / 结构顾问：CABR Technology Co. Ltd.
标志设计：Beijing Trycool Design / 室内CD公司：Choice Design / 景观CD公司：Miland Design
总承包商：Zhongxing Construction Co. Ltd. / 室内承包商：Jiangsu Construction Engineering Group, Jin Longteng Decoration Co. Ltd.
客户：Changyang Government of Fangshan District, Beijing / 功能：36 class junior and senior high school, classrooms, labs, auditorium, gymnasium, canteen, administration, dormitory and sport facilities / 用地面积：45,000m² / 有效楼层面积：57,773m² / 施工时间：2012—2014
摄影师：©Su Shengliang (courtesy of the architect) - p.74~75, p.77, p.78, p.81, 82^{top}, p.84, p.86^{bottom}, p.87^{middle} / ©Xia Zhi (courtesy of the architect) - p.73, p.85, p.86^{top}, p.87$^{top, bottom}$ / courtesy of the architect - 82^{bottom}, p.83

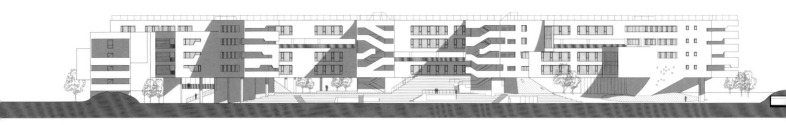

东立面 east elevation

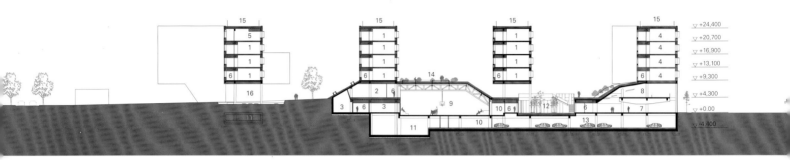

1 教室 2 音乐教室 3 工作室 4 实验室 5 图书馆 6 走廊 7 学生食堂 8 教工食堂
9 健身房 10 储藏室 11 机械设备间 12 竹园 13 停车场 14 花园 15 农场 16 池塘
1. classroom 2. music classroom 3. workshop 4. laboratory 5. library 6. corridor 7. student canteen 8. faculty canteen
9. gym 10. storage 11. mechanical room 12. bamboo garden 13. car parking 14. garden 15. farm 16. pond
A-A' 剖面图 section A-A'

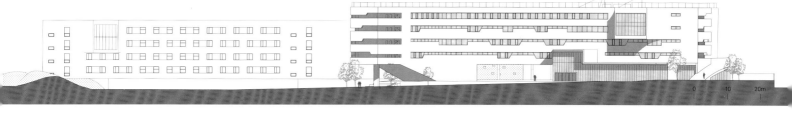

北立面 north elevation

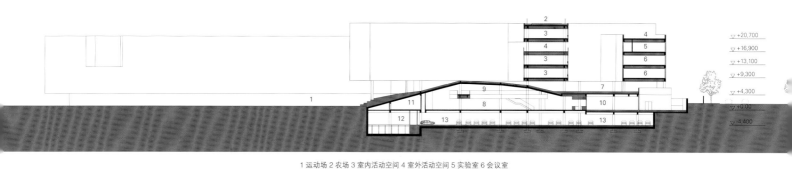

1 运动场 2 农场 3 室内活动空间 4 室外活动空间 5 实验室 6 会议室
7 花园 8 学生食堂 9 教工食堂 10 厨房 11 储藏室 12 机械设备间 13 停车场
1. sports field 2. farm 3. interior activity space 4. exterior activity space 5. laboratory 6. meeting room
7. garden 8. student canteen 9. faculty canteen 10. kitchen 11. storage 12. mechanical room 13. car parking
B-B' 剖面图 section B-B'

日压端子工厂

Ryuichi Ashizawa Architect & Associates

这座建筑是对原有工厂的扩建，它缓慢地融入周围的马来西亚Johor丛林。自19世纪以来，日压端子工厂就将理性和生产力作为工厂的主要动力，因此，建筑师的想法是改善工厂的工作环境，让工人为工厂的改变而自豪。

建筑师希望将碳足迹降到最低，更多地使用自然能源——雨水、太阳能、地热能和植被，从而使这座建筑成为可持续发展的工厂建筑。这个设计的目标是建造一个能盖住整座建筑主体的巨大绿色屋顶，屋顶一直延伸到地面，从而扩大了地表面积。建筑师从伊斯兰文化中特有的阿拉伯式花纹和周边的热带雨林风景中得到灵感，在建筑内部设计了顶端带有五角星的六边形柱子。

雨水可以顺着这些柱子内的封闭管道流到地下的蓄水池中，以备灌溉使用。建筑中主要的能源消耗是用于制冷，因此建筑师设计了一个水塘，有风吹进来的时候，水塘就可以降低室内温度。

为了能够减少电灯的能源消耗，工厂的天花板被设计成能够反射自然光线的效果。通过电脑模拟，建筑师预测了天花板所反射的自然光的亮度，并用有阿拉伯特色花纹的玻璃板将光线发散出去。

设计多层的办公楼，就要考虑到太阳辐射的问题。太阳光的热量会炙烤整座建筑的外墙，这也是建筑师将建筑设计成东西走向的原因。建筑外围的楼板作为斜坡连接地面，工人们可以每天在此锻炼身体，保持健康。这座办公楼的另一个设计特色是它的外表皮，立面是金属网格加固的藤蔓，整面的垂直绿植墙有效地挡住了太阳辐射。最后，废气从建筑内一座更高的建筑中排出，从而保证建筑下层空间的自然通风。

日压端子工厂的建筑理念就是要创造一个能融合在周边大自然中的工作空间。最终，它其实创造了一个室内自然环境。

Factory on the Earth

Kindly flowing into the surrounding Johor Jungle in Malaysia, this project is an extension of an existing factory. The factories in the 19th century were mainly driven by rationality and productivity. Therefore, we wanted to improve the factory working environment so that it's workers would be proud of it. Using the power of nature like rain, sunlight, wind, geothermal and vegetation, we wanted to minimize the carbon footprint, making the building a sustainable factory. The plan intends to create a large green roof continuous with the ground, extending the earth surface that would host the inside spaces. Derived from the Arabesque patterns of Islamic culture and as a reference to the surrounding jungle, the inside space is structurally designed by a forest of hexagonal-sectioned pillars with a star shaped top.

By means of enclosed pipes in the pillars, rain water is collected in an underground water storage tank. From here, the water is used for plants watering. As cooling the inside space

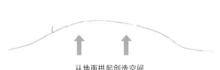
从地面拱起创造空间
putting up from the ground to make a space

用树木支撑地面
supporting the earth by trees

建筑体量落在底层地面上
giving the volume to the ground

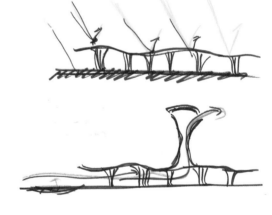

项目名称：Factory On The Earth
地点：Plot D04, jalan Tanjung A/4, Pelabuhan Tanjung Pelepas,
81560 Gelang Patah, Johor, Malaysia
建筑师：Ryuichi Ashizawa Architects / 当地建筑师：Nakano Construction sdn Bhd
结构工程师：Hirokazu Touki, Takuo Nagai
机械工程师：ES associates _ Eiji Sato / 电气工程师：Nichiei Architects
照明工程师：Izumi Okayasu / 照明设计：Izumi Okayasu
景观建筑师：WIN _ Junichi Inada / 客户：JST Malaysia
用地面积：46,489m² / 建筑面积：15,232m²
有效楼层面积：25,141m² / 景观面积：27,143m²
屋顶绿化面积：14,891m² / 建筑覆盖率：49% / 有效楼层比率：83%
设计时间：2010.12~2011.12 / 施工时间：2011.12~2013.5
摄影师：©Kaori Ichikawa (courtesy of the architect)

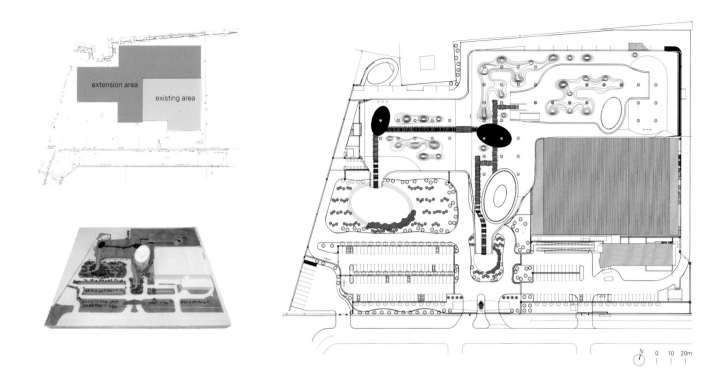

was one of the energy concerns, we designed a pond that would create cool air when wind blows.

In order to reduce the use of artificial light, the factory's ceiling is designed to reflect the light that comes from above. It was by computer simulations that we could predict the amount of skylight reflected and diffused by reflective Arabesqued-shaped panels.

As a multi-story office building, considering solar radiation was an important issue. That is why the building is aligned along the east-west direction while the heat is projected on the building's outer wall surface. The slab along the perimeter of the high-rise building functions as a slope that connects the ground level. Day by day, this continuous walking path is used by workers to practice exercise and improve their health. Another feature of this office building is it's outer skin. As the facade consists of a wired system of vines, this vertical green wall protects from solar radiation. Finally, air circulation is carried through a higher building within the building so that the lower spaces are naturally ventilated.

Creating a working place that is embodied with the natural surroundings of the site was the main concept that guided us throughout the design of the Factory on the Earth. In the end, it just became an inside natural environment.

Ryuichi Ashizawa Architect & Associates

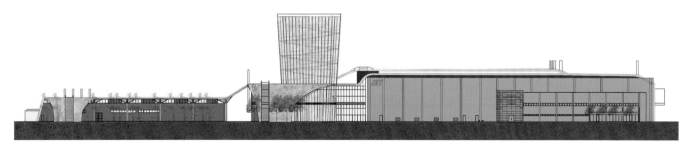

南立面 south elevation

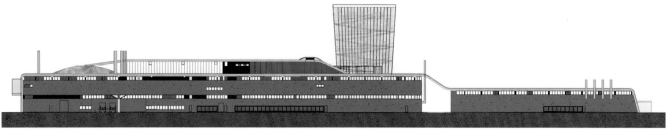

北立面 north elevation

西立面 west elevation

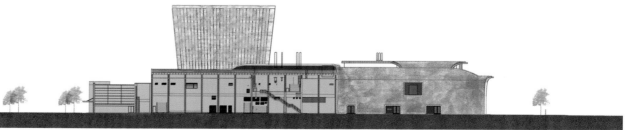

东立面 east elevation

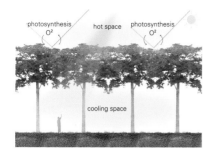
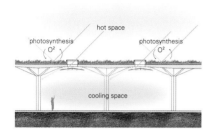

1 钎焊室	21 工厂办公室
2 调解室	22 隔离程序/剪辑储藏间
3 媒体室储藏间	23 办公室
4 男衣帽间	24 种植商店储藏间
5 女衣帽间	25 种植商店扩建区
6 鞋柜	26 展览区域
7 工人入口	27 庭院
8 统一媒体室	28 转角绿化
9 剪辑室	29 食堂
10 选择区域	30 厨房
11 仓库储藏间	31 女性祷告室
12 仓库	32 男性祷告室
13 包装储藏间	33 主席房间
14 庭院	34 图书馆
15 会议室	35 会议室
16 卸载区域	36 餐厅
17 仓库	37 更衣室
18 装货区域	38 澡堂
19 装货平台	39 室外澡堂
20 CNT选择/检查区域	

1. brazing shop	21. factory office
2. adjustment room	22. insulation process /cutting storage
3. press room storage	
4. male locker room	23. office
5. female locker room	24. planting shop storage
6. shoes locker	25. planting shop extension
7. worker entrance	
8. one off press room	26. exhibition area
9. cutting room	27. courtyard
10. selection area	28. corner green
11. warehouse storage	29. canteen
12. warehouse	30. kitchen
13. packing storage	31. female prayer
14. courtyard	32. male prayer
15. meeting room	33. chairman's room
16. unloading area	34. library
17. warehouse	35. meeting room
18. loading area	36. dining room
19. loading platform	37. dressing room
20. CNT selection /inspection area	38. bath house
	39. outside bath house

四层 third floor

五层 fourth floor

六层 fifth floor

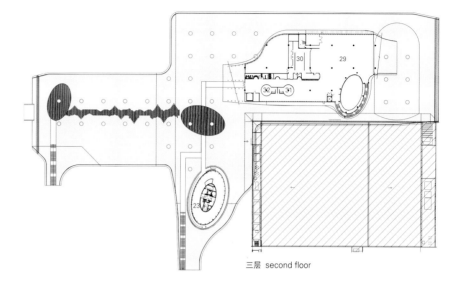
三层 second floor

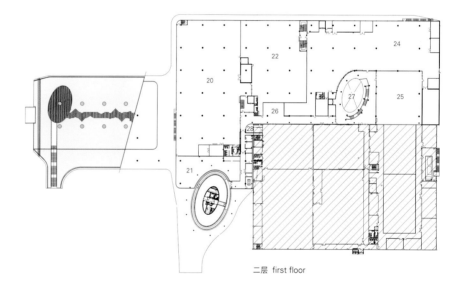
二层 first floor

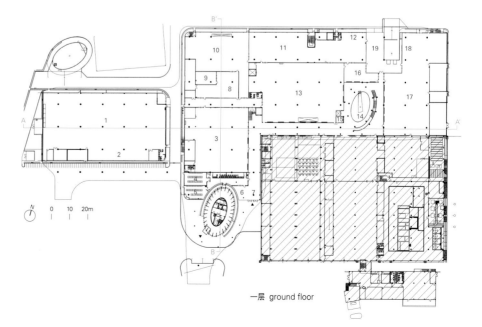
一层 ground floor

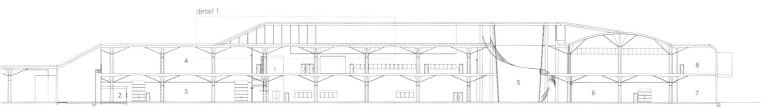

A-A' 剖面图 section A-A'

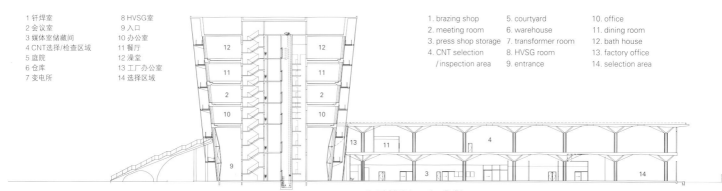

1 钎焊室	8 HVSG室	1. brazing shop / 5. courtyard / 10. office
2 会议室	9 入口	2. meeting room / 6. warehouse / 11. dining room
3 媒体室储藏间	10 办公室	3. press shop storage / 7. transformer room / 12. bath house
4 CNT选择/检查区域	11 餐厅	4. CNT selection /inspection area / 8. HVSG room / 13. factory office
5 庭院	12 澡堂	9. entrance / 14. selection area
6 仓库	13 工厂办公室	
7 变电所	14 选择区域	

B-B' 剖面图 section B-B'

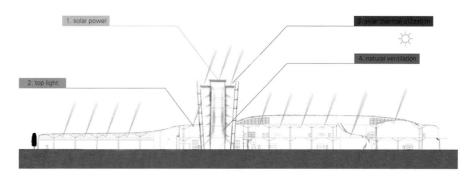

1. It reduces the electricity used by factory and generates electricity using sun light.
2. Considerates the direct light, and incorporates light in the factory.
3. By using the solar heat, and then used for hot water supply of the public baths.
4. Make the natural ventilation using the height of the high-rise building.
5. The light is shielded by eaves.
6. The insulation effect of the green roof.
7. Wind that was cooled by the heat of vaporization of biotop will be sent to architecture.
8. Rainwater is reused to water the roof.

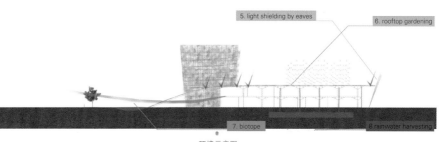

环境示意图
environment diagram

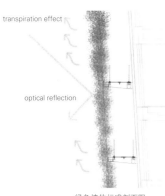

绿色墙体标准剖面图
green wall typical section

Bauhinia Kockiana

congea tomentosa

epipremnum pinnatum

piper nigrum

thunbergia laurifolia

thunbergia grandiflora

thunbergia laurifolia 'alba'

momordica charantia

绿色墙体植被
green wall vegetation

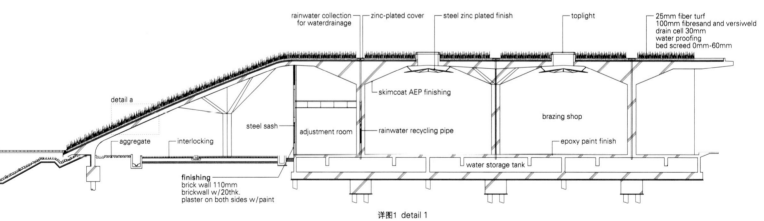

详图1 detail 1

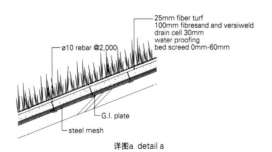

详图a detail a

能源意识与可持续公共空间 Energy-conscious and Sustainable Public Spaces

Clos des Fées村
Mutabilis Paysage + CoBe Architecture

Clos des Fées村位于距离Paluel峭壁几百米的地方，距离核电站不是很远，是一个由Conteville村延伸而来的区域。Clos des Fées村包括八个分布在传统庭院周围的独立房间，两个公寓和三个艺术家工作室。设计委员会将注意力主要集中在西面房间的布局上，强调真正的放松空间，同时避免城市衰退。

Clos des Fées村迎着海风，也经常隐蔽在细雨蒙蒙之中。

这个项目充分利用了这些条件，并在山谷中修建了连通的山谷网络，这些网络更多的作用是作为水管理工具和天然的框架，不再需要栅栏，同时也增强了童话公园的吸引力。

风力也是整个能源项目的驱动力，风力农场给水池、人工湖和水景园周围的水力系统提供了动力。涡轮机连接到照明雕塑上，雕塑的亮度随风力的大小而变化。

Le Clos des Fées Village

Located a few hundred meters from the cliffs of Paluel, not far from the nuclear power plant, the Clos des Fées is a proposed extension of the village of Conteville. Curated by an iconic team, the communal housing, and the Garden of Fairies, the Clos des Fées consist of 8 individual houses organized around traditional courtyards, 2 vacation rentals, and 3 artists' studios. The planning committee focused on concentrating the housing density in the western part of the site so as to allow for true breathing spaces and avoid urban decay.

A-A' 剖面图——社区空间
section A-A'_ community space

B-B' 剖面图——社区空间
section B-B'_ community space

C-C' 剖面图——社区空间
section C-C'_ community space

D-D' 剖面图 section D-D'

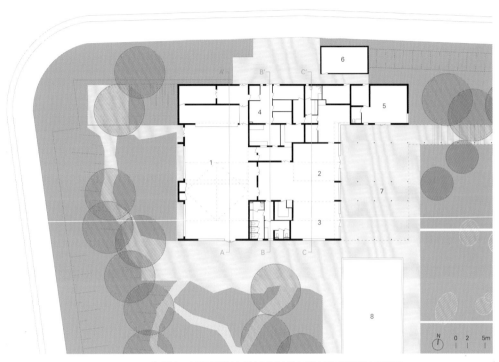

1 多功能室 2 大厅 3 冬季花园 4 厨房 5 商业空间 6 当地公园 7 有顶庭院 8 童话公园
1. multipurpose room 2. hall 3. winter garden 4. kitchen 5. commerce 6. local park 7. covered courtyard 8. the fairy Basin

一层——社区空间
ground floor_ community space

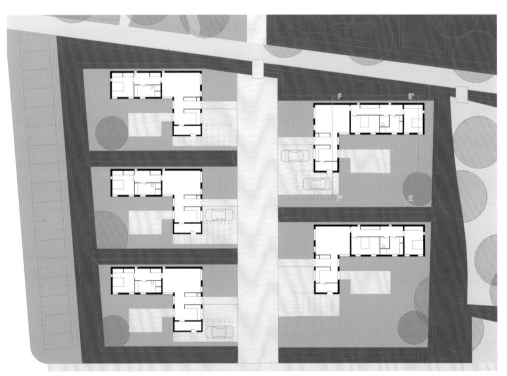

一层——住宅
ground floor_dwelling

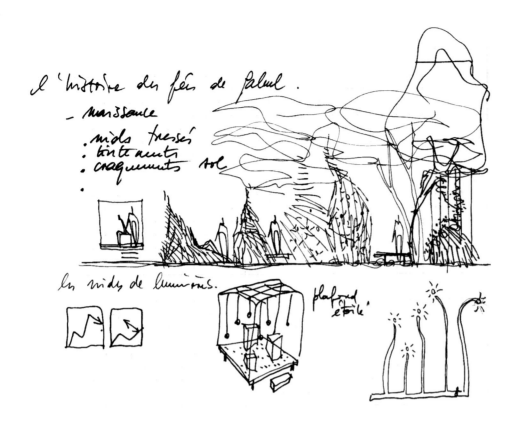

The site is buffeted by winds coming from the sea, and is often shrouded in a drizzly mist.

The project makes the most of these conditions, however, by establishing a network of valleys that are as much a tool for water management, a natural framing device that eliminates the need for fences, and an attraction in the fairy garden.

The wind is also the driving force of an energy project. The establishment of a wind farm activates the hydraulic system developed around basins, sheets of water, and water gardens. The turbines are connected to light sculptures whose brightness vary with the force of the wind.

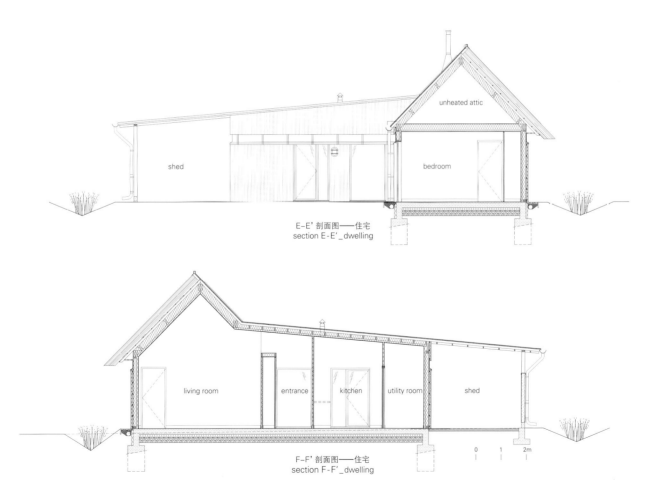

E-E' 剖面图——住宅
section E-E'_dwelling

F-F' 剖面图——住宅
section F-F'_dwelling

项目名称：Le Clos des Fées / 地点：Paluel, France / 建筑师：Mutabilis Paysage, CoBe Architecture
项目团队：Ville de Paluel, CoBe Architecture, Mutabilis Paysage / 用地面积：56,700m² / 建筑面积：467m²
造价：EUR 7,331,000 / 设计时间：2007 / 施工时间：2012 / 竣工时间：2013 / 摄影师：©Luc Boegly (courtesy of the architect)

2015世博会：绿色实验室
EXPO 2015: The

尽管人们对举办世博会的实效性和对2015米兰世博会目前存在的问题仍有争议，但是本届世博会的主题和主线"给养地球，生命的能源"颇具吸引力。

健康食物是指对世界来说既安全又充足的食物。2015米兰世博会深入彻底地反映了保证全世界拥有健康食物的策略，成为创新和研究的实验室，同时也为进行国际讨论和辩论提供了契机，讨论和辩论的主题是如何在尊重地球和保证其平衡的同时，为全世界提供健康、安全和充足的食物。

整个展馆区有一百多万平方米，分布在两个正交轴上，分别是Cardo（一条大约400m长的轴线）和Decumanus（一条大约1.5km长的轴线），反映了古罗马人的城市规划风格。运河水道环绕四周，通往各展馆的路线上有大型的雨篷遮盖，为游客遮阳挡雨。意大利馆位于Cardo轴线上，其他140个展馆大多沿Decumanus轴线分布。为了建造节能、可持续的建筑物，所有展馆都可以在活动结束后拆除并再次使用。

散落在意大利馆周围的组合馆看起来特别吸引人。每个组合馆由一些有相同食物背景（例如，大米、咖啡和香料）的国家或者其他相同背景（例如干旱地区和岛屿）的国家组成。

Marco Atzoril深入研究了世博会。

Thematic Areas	13. Lithuania	27. European Union	43. Morocco	59. Nepal	71. Caritas Internationalis	85. New Holland Agriculture
1. Pavilion Zero	14. Belarus	28. Switzerland	44. Iran	60. Ireland	72. Ven. Fabbr. del Duomo	86. Coca-Cola
2. Expo Centre	15. Malaysia	29. Ecuador	45. Chile		73. Save the Children	87. Expo Partner 1
3. Children Park	16. Thailand	30. Germany	46. Austria	Clusters	74. Cascina Triulza	88. Alitalia Etihad
4. Biodiversity Park B. Fiere	17. Uruguay	31. Kuwait	47. Slovenia	61. Rice	75. Don Bosco Network	89. China Corporate Un. Pav.
5. Biodiversity Square Slow Food	18. China	32. USA	48. Mexico	62. Cacao and Chocolate	76. World Ass of Agronomist	90. USA Food Truck Nation
6. Future Food Dist. Coop	19. Colombia	33. Turkey	49. Romania	63. Coffee		91. Alessandro Rosso-Joomoo
	20. Argentina	34. Monaco	50. Spain	64. Cereals and Tubers	Companies	92. Expo Partner 2
National Pavilions	21. Poland	35. Japan	51. Hungary	65. Arid Zones	77. Casa Algida	93. Federalimentare
7. Czech Republic	22. Netherlands	36. Slovakia	52. UK	66. Islands, Sea and Food	78. Enel	94. McDonald's
8. Bahrain	23. Holy See	37. Russia	53. Kazakhstan	67. Bio-Mediterraneum	79. Technogym	
9. Angola	24. France	38. Estonia	54. UAE	68. Spices	80. Perugina	Events Areas
10. Brazil	25. Israel	39. Oman	55. Azerbaijan	69. Fruits and Legumes	81. Lindt/Distretti Cioccolato	95. Conference Centre
11. South Korea	26. Italy	40. Indonesia	56. Vietnam		82. Franciacorta	96. Auditorium
12. Moldova	a. Palazzo Italia	41. Turkmenistan	57. Belgium	Civil Societies	83. Intesa Sanpaolo	97. Open Air Theatre
	b. Copagri	42. Qatar	58. Sudan	70. Kip - International School	84. Vanke	98. Lake Arena

Green Laboratory

In spite of the controversies not only over Expo validity but also the current Expo Milano 2015 pending issues, the main theme and the common thread of this Expo, "Feeding the Planet, Energy for Life", is appealing.

With serious and thorough reflection on strategies to ensure healthy food that is safe and enough for the world, Expo Milano 2015 opened a laboratory of innovation and research, a time for international discussion and debate to guarantee healthy, safe and sufficient food for the world, while respecting the Planet and its equilibrium.

The exhibition area is more than one million square meters and is spread over two orthogonal axes, reflecting the signature urban-planning style of the ancient Romans, called Cardo (on an axis of about 400 meters) and Decumanus (about 1.5 km long). The site is completely surrounded by a canal and large canopies placed on the routes, helping visitors shelter from the rain and the sun. The Italy Pavilion is located along the Cardo, while the most of the other 140 pavilions are lined along the Decumanus. Constructed to be energy-efficient and sustainable, the buildings were designed to be easily removed and reused after the event concludes.

The Clusters located scattered around the Italy Pavilion particularly look attractive. Each cluster is shared by multiple countries with common food background (for example rice, coffee and spices, or arid zones and islands).

Marco Atzori looks at the Expo in depth.

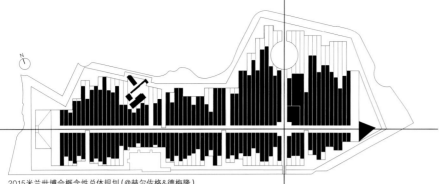

2015米兰世博会概念性总体规划（@赫尔佐格&德梅隆）
conceptual masterplan of Expo Milano 2015 (©Herzog & de Meuron)

场地鸟瞰 panoramic view of the site

西侧鸟瞰 aerial view from the west

东侧鸟瞰 aerial view from the east

2015世博会：绿色实验室

Marco Atzori

　　2015年的米兰世博会代表了迄今为止建筑方面最先进的研究成果。此外，它标志着与可以代表最近几年建筑特点的最新模型和模式的距离。因为世博会展出的作品都堪为建筑杰作，所以说2015世博会的展馆向人们发出了一个极其有趣的信号。每一个设计项目都表现出一个共同特征，那就是它们都是材料、工艺和设计方法的综合体。最先进的设计方法和建造工艺的结合使用表明了人们对建筑可持续性的不断关注。可回收材料的使用、可拆除的结构、对干燥系统的关注和使用轻质预制组件的诸多优势，都标志着人们对资源平衡问题的广泛关注，同时也表明了人们对当代设计和建造过程的成熟处理。材料方面的技术运用，比如，钢结构和层压木材，进一步完善了轻质体系的潜力。生产体系和组装已经可以数字化控制，使复杂形状的生产成为可能，使用量越来越大，也备受设计师的关注。近年来，经济危机对欧洲多个国家（包括主办国意大利）的经济和社会产生了影响，人们努力寻求节能、环保、高效的建筑方法，2015世博会上许多有建设性的、有意义的选择都受到了这一趋势的影响。

　　2015世博会为未来打开了一个窗口，主要关注2007年至2012年全球经济危机以后出现的问题，在这里人们看到了让人乐观的新迹象。此外，建筑的可拆装、建筑的规格大小、建筑方法的选择和世博会的

EXPO 2015: The Green Laboratory

Marco Atzori

The 2015 Universal Exhibition in Milano is an indicator of the most advanced research in architecture to date. Moreover, it marks a distance from the final models and forms that have characterized the architecture of recent years. As the Exhibition displays works that can be considered masterpieces of architecture, the Expo 2015 pavilions offer signals that extremely interesting to consider. A common aspect emerging from the presentation of each project has been the tendency to the synthesis of materials, technics and design methodologies. As result of the combination of the most advanced design methodologies and construction techniques, a growing focus on sustainability has been shown. The use of recyclable materials, dismantled structures, attention to dry systems and the advantages of a light prefabrication mark a widespread awareness of the question to the balance of resources and indicate a mature way to manage the contemporary design and construction process. The use of technology applied to materials has led to the refinement of the potential offered by light systems, such as steel structures and laminated wood. The digital control of the production systems and assembly has made possible the development of complex shapes used with greater capacity and attention by designers. In Expo 2015, constructive and meaningful choices are influenced by the search for a new efficiency of architecture related to the derivations of the economic crises that have affected the economy and society of several countries in Europe (including the hosting Italy) in recent years.

Cardo轴线 Cardo

Decumanus轴线 Decumanus

开幕式 opening ceremony

湖边活动场所 lake arena

主题都表明一种世界经济全新的政治地理环境。并非巧合的是,与西方巨头,比如美国和德国相比,本届世博会最大而又最先进的展馆是韩国馆、巴西馆、中国馆、安哥拉馆以及哈萨克斯坦馆。这些新成长的经济体利用世博会公开肯定自己在世界舞台上的地位。

虽然可以从总体上对2015世博会进行理解,但是每个展馆都有着不同的特点与展示内容,必须有多样的解读方式。大多数展馆都被看作是单一的、临时性的建筑,其内在体验和外在体验明显不同。

整个世博会的基础设施基于Cardo和Decumanus这两个概念,它们是古罗马城市的两条主轴。所选的网状结构对游客从内部街区观看单一场馆有巨大的影响。网格的使用给人们的整体印象是世博会是由单一突出的展馆组成的,而不是统一城市规划的展现。给人这一离散的印象实际上偏离了策划团队最初想要传递给人们的理念:整个展区就是一个大型的城市花园,在这里,所有建筑遵循共同规则,通过自然元素密切连为一体。然而,打造大型城市花园的想法只在几个展馆设计中得以实现,整个看起来根本没有最初规划理念的踪影,是对最初规划理念最保守的执行。

奥地利馆——奥地利馆被设计成一座人工森林,"呼吸"是它的主

Expo 2015 can be considered a doorway that opens to the future, because it focuses on possible scenarios after the 2007–2012 global financial crisis, in which there seem to be new signs of optimism. Furthermore, displacement, size, construction choices, and expositive themes show a new political geography of the world economy. It is no coincidence that, compared with Western giants like the United States and Germany, the largest and most advanced pavilions are Korean, Brazilian, Chinese, Angolan and Kazakhstani. The new growing economies use the Expo to affirm definitively their presence on the world stage.

Although a general reading for the Expo 2015 is possible, the nature and the contents offered by each pavilion are so diverse, that multiple codes of reading should be used. Most of the pavilions have been conceived as singular and temporary objects, with a clear cut between the inside and the outside experience.

The basic infrastructure of the overall Expo is based on the concept of the Cardo and Decumanus, the main axes in Ancient Roman cities. The chosen grid structure has a great impact on the visitors' perception of the single pavilions from the internal streets of the event. The overall image resulting from the application of the grid is that of a collection of standing-out objects, rather than the manifestation of a unified plan. This discrete resulting image diverges from the expectations that the planning team elaborated in the initial concept. The site had to be a large urban garden, in which architectural objects were controlled by common rules that were intended to intimate a connection with the natural elements. However, the idea of the large urban garden has been

题。展馆里有人工创造的大自然,其目的是使人类与微观宇宙之间产生可能的相互作用。微观宇宙的诞生、成长、自我繁殖以及死亡都在这个环境中发生。氧气和光合作用是人类与大自然之间紧密关系的重要体现,这种联系强大、有力但却无形,是一种可被用于许多形式、用于不同目的的能量。展馆的突出之处是内外世界之间明确的划分:外部是世博会拥挤的空间,各展馆之间游客人头攒动;内部是弥漫在将游客与之隔离开来的内部空间中的大气,使人能体验到一种既亲密熟悉又非常独特的氛围。展馆被设计成一个悬挂于地面之上的空盒子,四周密不透风,没有开口,由七个灰色不透明混凝土表面围成。这一设计强调内外世界之间明确的划分,使游客渐渐融于令人倍感亲切的内部空间。按照建筑师的说法,这一迷人的设计旨在向游客介绍本质上能代表奥地利领土的东西——自我隔离,这与彼得·卒姆托在他的蛇形画廊设计中所营造的氛围类似。

巴西馆——当今的建筑趋势是通过结构外观来体现建筑特色,而巴西馆是巴西建筑现代矩阵与当今建筑趋势两者之间一个有趣的综合体,在建筑结构外部和内部之间建立了富有成效和充满智慧的联系,使巴西馆融于整个世博会的构架之中。空间试验和建筑语言从不同层面把巴西馆定义为2015世博会最好的展馆之一。首先,你可以感受到基于对立的双方之间的对话:结构和体量之间,材料和材质之间,

followed only in isolated cases, and what appears is a depleted and most conservative development of initial concepts.

Austria Pavilion – The Austria Pavilion is designed as an artificial forest, in which "breathe" is the main theme. The pavilion possesses an artificially created nature, the aim of which is to enable possible interactions between humans and the microcosmic universe that is born, grows, reproduces itself and dies in the environment. Oxygen and photosynthesis are key representatives of the primal bond between man and nature, a connection that is strong and powerful but invisible, an energy that can be used in many forms and for different purposes. The strength of the pavilion is determined by the clear separation between the outside and the inside worlds, between the crowded space of the Expo, swarming with visitors moving between the pavilions, and the atmosphere enveloping the interior space that isolates the visitors, offering an intimate and almost individual atmosphere. The pavilion is designed as an empty box suspended above the ground and its impenetrable perimeter, created by seven blind gray concrete surfaces, without openings. This structure underlines the fracture required for immersion into the intimate space represented in it. A fascinating choice is intended to introduce the visitors to what, in the view of the architects, becomes to essentially represent the Austrian territory that stages a self-isolation similar to the atmosphere created by Peter Zumthor in his design of the Serpentine Pavilion.

Brazil Pavilion – The Brazil Pavilion is an interesting synthesis between the modern matrix of Brazilian architecture and the current tendency to characterize the architectural work

内外氛围之间。其次,在建筑师的设计之下,通过使用大型的弹性网状结构,使游客的体验具体化。这个拉伸的网状通道通向内部空间的入口,是整个场馆最显著的特点,游客的行走过程从而变成了一种在充满遐想的空间里的游戏体验。室外区域没有奢侈铺张,而是被重新诠释成一种感官体验,人们可以闻到植被的气味,可以听到人工制造的、让人想到亚马逊森林的声音。

意大利馆——意大利馆是在现场施工建造的,给人印象最为深刻,是在世博会结束之后唯一有可能重复使用的展馆,也是唯一使用重型建筑技术、用钢筋和混凝土建造的展馆。这座建筑仅仅通过抽象的东西——包围在建筑外面的外表皮来反映自然元素,建筑语言没有对环境或本次世博会的统一主题做出任何让步。这是由Nemesis及合伙人事务所建造的首座大型建筑,与当今建筑语言的发展相比,更加接近最近十年早期的研究实验。复杂几何图形以及在建筑立面进行的实验与某些美国建筑有着千丝万缕的联系,但并没有将Morphosis事务所的设计作为参考。意大利馆占据整个地块周边,设计有多孔的实心网状表皮,矗立在一个广场上,屋顶为玻璃屋顶,可以将雨水疏导到底部大型的漏斗中,把水排到广场内,营造出很棒的视觉效果。

韩国馆——韩国人心中的展览应该是在一个大大的白色室内空间

through the use of structural skins. There is a productive and smart relationship between the exterior and interior that focused on the concept of this pavilion within the overall structure of this Expo. Several levels of experimentation of space and architectural language help to define the Brazil pavilion as one of the best of Expo 2015. In the first level, you can sense a dialogue between parties based on the opposition between structures and volumes, between materials and textures, between atmospheres perceived of interior and exterior. In the second level, the visitors' experience intended by the architects is materialized by means of the large elastic net that leads to the entrance of the internal volume, one of the most striking feature of the entire pavilion. The visitors' path is thus transformed into a gaming experience that is introduced in a more contemplative space. Instead of outgoing extravagance, the outdoor area is reinterpreted to become a sensory experience linked with the scents of vegetation and sounds artificially produced that recall the Amazon forest.

Italy Pavilion – The Italy Pavilion is the most impressive structure built on the site and the only one that will probably reused once the event moves away. It is the only building in which prevails a heavy construction technology based on the combined use of reinforced concrete and steelwork. The building recalls the natural element only through its abstraction represented by the skin that surrounds it and is developed through a language without any concession to the context or unifying theme of the event. The work, which is the first big structure built by Nemesis & Partners, is part of a trend closer to the experiments of the early years of this decade than the current developments in the architectural

>>零号馆 _ Architetto Michele De Lucchi S.r.l.

从西入口进入2015米兰世博会,首先映入游客眼帘的是零号馆。零号馆旨在介绍"给养地球,生命的能源"这一世博会主题,告诉参观者这一主题中所蕴含的丰富内涵。零号馆被构思成地球地壳的一部分,造型整洁灵巧,激发人们继续探索我们的星球及其不为人知的秘密。整个场馆分为十个展厅,展厅内光线较暗,只有展品脱颖而出,清晰可见。每个展厅专门突出展示一个不同的人类文化产品。这个矩形的展馆占地大约10 000m²。从建筑的角度看,展馆使用云杉木板,有的地方层层排列,就像地图上的等高线,用来表示地壳表面的地形地貌,有山脉,有丘陵,有宽阔的中央山谷。这里也有盘古大陆,用许多桌子摆成拼图一样的组合。盘古大陆是向人们呈现当陆地、海洋、人类和大自然混沌一体时地球看起来也许是什么样子。

>> Zero Pavilion _ Architetto Michele De Lucchi S.r.l.

Zero Pavilion is the first to greet visitors arriving at Expo Milano 2015 from the west entries. Assigned to introduce the theme of "Feeding the Planet – Energy for Life", the pavilion tells visitors about the wealth of aspects inherent in this title.

It is conceived as a portion of the earth's crust, neatly cut and raised as an invitation to delve into our knowledge of the planet and its secrets. Reproduced inside the pavilion are ten caves in a semi-darkness against which the exhibits stand out clearly. Every area is dedicated to a different human cultural product. The rectangular pavilion occupies an area of about 10,000 sqm. From the constructional point of view, it consists of spruce board and partly practicable tiers, resembling the schematized contours used to indicate ground reliefs. The stratification of curves reproduces the earth's crust, with mountains, hills and a broad central valley. It is also host to Pangea, with multiple tables forming a puzzle-shaped combination. Pangea is a reconstruction of what the planet might have looked like when lands, oceans, men and nature were all one.

举行,形式复杂多变。只要非常精致和体现一定的设计理念,任何内部空间都可服务于基于传统和创新相结合的展览主题。"人如其食"这句格言充分体现了韩国馆"营养"这一主题。为了迎合这一理念,凭借空间建构和媒体技术所提供的支持之间强大的协同作用,韩国馆的展览给人以深层次的认知体验。

从这个意义上来说,韩国馆中现实和数字媒体的结合很好地展示了2015世博会举办的目的——能够实现你未来希望拥有的文化体验,是最好的展示之一。韩国馆是本次盛会规模最大和参观人数最多的展馆之一,其建筑和技术的运用都充分证明了韩国近年来政治和经济的改变,证明了韩国经济的增长。

中国馆——中国馆的设计如同"麦浪"覆盖全馆。展馆屋顶采用了以胶合木为主材,以竹层压木作为木梁的结构。馆内再现了中国传统农业平原,物景交融,空间自然过渡。馆内馆外具有连续性。无论从规模还是所代表的国家来讲,中国馆无疑都是2015世博会最重要的场馆之一。与其他诸如韩国馆、巴西馆、安哥拉馆、泰国馆一样,中国馆为2015世博会呈现了世界新的地理政治平衡。实际上,这些建筑物是让人们感知当代经济政治中心转移的信号,向西方体系定义了东方体系。另外,以中国馆建造设计为例,建筑技术注重轻盈轻便,以方便世博会前组装及世博会后拆卸。所有层压木板、覆盖屋顶的竹编材料和其他材料,都因为有了当前的生产和现场组装技术,才有可能在短时间内组

language. The use of complex geometries and experiments on the skin facade is linked to what is produced by certain American architecture than has the Morphosis as a point of reference. The building occupies the lot for the entire perimeter and is designed as a porous solid suspended in a square covered by a glass roof that has the peculiarity of channeling rain to a large funnel and then drains the water inside the square creating a great visual effect.

Korea Pavilion – The exhibition conceived by Korea is in a large and white interior characterized by a complex form. Highly refined and conceptual, any interior space is at the service of the exhibition theme based on the juxtaposition of tradition and innovation. The subject of nutrition in the Korea Pavilion is fully represented by the motto "You are What You Eat". Playing this concept, the exhibition develops through a powerful synergy between the construction of the space and the support given by the media technology, which creates a deep cognitive experience.

In this sense, the combination of reality and digital media in the Korea Pavilion is one of the best demonstrations that the Expo 2015 is able to achieve the cultural experience that you expect to live. The Korea Pavilion, one of the largest and most visited of the event, bears witness to the political and economic shifts of recent years and the growth of the Korean economy represented so powerfully by the architectural and technological choices used.

China Pavilion – The China Pavilion is designed as a great cover that is reminiscent of a wave created by a system of panels laid on bamboo laminated wooden beams. The cover is suspended over a landscape that recreates the traditional

装完成。从这一点来看,如何从技术控制、效率以及可持续性方面来重新考虑建造施工系统给人们带来的可能性,中国馆就是一个很好的实例。

英国馆——英格兰做出的展馆设计选择与2010上海世博会英国馆的设计如出一辙,都是一个非常壮观的、参观者可与之交互的雕塑。场馆由Wolfgang Buttress与结构工程师Tristan Simmonds合作设计完成,灵感是蜂巢和蜜蜂的生命周期。该场馆由漂亮的不锈钢格栅建造,参观该展馆的游客首先穿过花园,花园被抬高,游客看花园的视线高度就像昆虫在花园飞动的高度一样。然后,游客穿过通过感知和技术营造的场景,继而进入展馆主体——蜂巢。经过数字化处理的数据来自于位于英格兰的一个真正的蜂巢,随后被转变成光脉冲。Buttress和Simmonds所使用的建筑语言借鉴了英国建筑历史上最具试验性时期的设计,从巴克敏斯特·富勒的圆顶建筑到赛德克·普莱斯的结构方法,而控制学又将他们的设计应用于当今时代:如今,要保证生命的可能性和地球的发展,一定要达到建筑技巧和自然的环境平衡。英国馆的声音环境同样具有试验性,音乐由Sigur Ros(冰岛的后摇滚乐团)作曲,该展馆与世博会的其他展馆不同,没有封闭的部分,也没有显示任何信息,但其本身就是一个信息体。这个场馆在世博会上是独一无二的。

Chinese agricultural plain and produces a fusion between object and context that leads smoothly into the space. There is continuity between outside and inside. It is one of the most important pavilions of Expo 2015 both in size and for the nation that it represents. Along with other pavilions, such as, for example, Korea, Brazil, Angola and Thailand, the China Pavilion presents the new geopolitical balance of the world into the Expo 2015. These buildings are in fact the signal perception of the shift of the contemporary economic and political centers of gravity that define the Eastern counterpart to the Western system. In addition, in this case, the construction techniques have focused on lightweight systems that can be easily assembled and then dismantled according to the strategies for post-expo. The elements in laminated wood, the bamboo panels for covering and all other systems were assembled in record time thanks to the possibilities offered by current systems of production and assembly on site. From this standpoint, the China Pavilion is a tangible example of the possibilities of rethinking the construction system in terms of technical control, efficiency and sustainability.

UK Pavilion – England makes a choice in line with what was presented in Expo Shanghai 2010 preferring to present its pavilion as a spectacular interactive sculpture designed by Wolfgang Buttress in collaboration with the structural engineer Tristan Simmonds based on the idea of the hive and the life cycle of bees. The building is made with a beautiful stainless steel system. It is reached through a garden leveled to be observed at a height reached by insects in flight. Visitors pass through a sensorial and technological scenario that leads up to the dome that is the hive. This digitally reprocessed data

  法国馆——法国馆向人们揭示了当今参数控制技术在设计建造复杂几何形状中应用的无限可能，也向人们展示了如胶合板这样取材于自然原材料的科技材料应用的潜力。该场馆被设计成耕地景观，生长在法国这片疆域上，代表着法国农业的不同植物物种。内部空间就像是一个洞穴，由复杂的木网格结构组成，这是通过参数控制系统切掉胶合板中不同的元素来实现的。其结果叹为观止，内部空间非常敏感，充满刺激。

  法国馆就是一个"大市场"，弧形的屋顶上挂满了产品，部分代表了经济技术合作趋势。这种趋势在一些法国文化作品中已经有所体现，早期的实验者包括Francois Roche及他的工作室R & Sie(n)。法国馆的设计通过利用混合材料、数字技术和自然元素，强调景观与建筑作品的融合。通过数字工具控制材料的独特建筑策略在法国建筑师的作品中无疑是成熟的，尽管受到场地建筑密度的影响，但是其结果仍然是显著的。

  巴林国馆——这是一个秘密花园，是一片待人们去发现的绿洲，隐藏在白水泥板围成的流动空间里。设计师对前缩透视法和运动空间的灵活掌握，使白水泥板在这样一个空间内不时地定义出不同的视角，将空间放大，使其形式多样。世博会巴林国馆由安娜·霍拖普设计，她重新思考了阿拉伯花园的设计方法，用白色混凝土墙对花园加以

from a real hive located in England is transformed into light pulses. The language adopted by Buttress and Simmonds refers to the most experimental period in the history of English architecture from the geodesic domes of Buckminster Fuller to Cedric Price's approach to the structure and cybernetics transferring them in an era when the environmental balance between artifice and nature must be reached to keep the possibility of life and development for the planet. The sound environment, equally experimental, is a Sigur Ros composition. The pavilion, unlike all the others in the Expo, has no closed parts, does not expose information, but is an informational structure itself. It is unique in the Expo.

France Pavilion – The France Pavilion reveals the possibilities offered by today's technology of parametric control in the development of complex geometries and the potential offered by technological materials from materials of natural origin, such as laminated wood. The pavilion also features an agrarian landscape that hosts different plant species typical of the territory and of French agriculture and introduces a space dug like a cave. The shape is achieved through a complex network structure by parametric controlled systems cutting different elements of laminated wood. The result obtained is remarkable, and the interior space is sensitive and full of stimuli. Designed as a market where the products are shown on the curves of the roof, the France Pavilion forms part of an ecotech trend present in some cultural French context among whose early experimenters include Francois Roche and his studio R & Sie(n). The pavilion emphasizes the fusion between landscape and architectural objects obtained through the use of hybrid materials, digital technology and natural elements.

>>组合馆

组合馆是2015米兰世博会最具创新性的元素之一。米兰世博会上这些组合馆超出了往届世博会传统的按照地理位置组合的联合馆。事实上，组合馆是把生产同类食物的国家或有兴趣开发同样具有代表性主题的国家聚合在一个建筑项目下。组合馆模式鼓励相距甚远的国家彼此开展文化与传统的对话，分享在农业、营养、福利和可持续发展领域可能的应对相同挑战的办法。

2015米兰世博会有9个组合馆：6个按食物链（咖啡、水稻、可可豆和巧克力、香料、水果和蔬菜、谷类和薯类）划分，其他3个按与营养有关的特定主题（生物地中海、干旱地区的农业与营养、岛屿、海洋和食物）划分。

>> Clusters

The Clusters represent one of the most innovative elements introduced by Expo Milano 2015. These exhibition spaces go beyond the traditional Joint Pavilions grouped on a geographical basis of the passed Expos. In fact, their aim is to join under the same architectural project. Countries that have in common the production of the same food product or that are interested in developing a shared and representative theme. The Cluster model invites countries to a dialogue among cultures and traditions even distant among each other, and to share possible solutions to common challenges in the sectors of agriculture, nutrition, welfare and sustainable development.

Expo Milano 2015 has developed nine Clusters: six are dedicated to food chains (Coffee, Rice, Cacao and Chocolate, Spices, Fruits and Legumes, Cereals and Tubers), while three are dedicated to specific themes connected to nutrition (Bio-Mediterraneum, Agriculture and Nutrition in the Arid Zones, Islands, Sea and Food).

- Rice _ Growth Reflections
- Cacao and Chocolate _ Unfolding Flavour
- Coffee _ Coffee in the Forest
- Fruits and Legumes _ United Support
- Spices _ Spice Voyage
- Cereals and Tubers _ The Valley
- Bio-Mediterraneum _ The White Dream
- Islands, Sea and Food _ Bamboo Forest
- Agriculture and Nutrition in the Arid Zones _ Dustorm

分隔围挡，用一些基本要素突出种植在天井里的植物，与外界完全隔离开来。这是一个可供人静观和打坐的空间。

德国馆——德国馆最引人注目的特点就是有一个大广场，游客可以直接从广场走到一个小型露天剧场，这里上演着不同的节目。屋顶上的广场由坡道组成，坡道连通到地面，这里是一个公共项目所在，因此划分了室内室外展览空间。德国馆和巴西馆都选择在公共空间传递其使用体验，在避免成为一座自我指涉的建筑的同时，创建了一个半公共的维度，打破了主要基础设施（公共开放空间）和场馆（封闭的私密空间）之间的二元对立。尽管这样的诠释早在2015世博会城市概念的初稿中已被提议，但是由于种种原因仍然没有实现。这样的设计已经没有了建造全球性大花园的想法，而是回归到传统的城市空间建设。在传统的城市空间里，街道和广场代表着公共空间，而场馆则可容纳各种活动。

只有德国、巴西、法国，还有从较小的程度上说中国和韩国，已经以不同形式在场馆设计中形成了这种可能的互动，这在世博会的主题背景下，可能是重要的试验场。

阿联酋馆——阿拉伯联合酋长国委托诺曼·福斯特负责其场馆设计。诺曼·福斯特是本届世博会上少有的知名设计师之一，与这个中东

The peculiar construction strategies of controlling materials through digital tools turn out to be mature in the work of French architects' work, and the result, although affected by the density of occupation of the site, is remarkable.

Bahrain Pavilion – This is a secret garden, an oasis to discover inside a fluid space created through the use of baffles in white cement that define from time to time different perspectives within a space that as concluded, thanks to the mastery of foreshortening and of the movement, is amplified and multiplied. The exhibition Expo Bahrain is thanks to the work of Anne Holtrop, who rethinks the idea of an Arab garden bounded by white concrete partitions. A few essential elements highlight the vegetation hosted inside patios in a space totally separated from the outside. The space is meditative and contemplative.

Germany Pavilion – The most intriguing feature of the Germany Pavilion is the large square/cover walkable by visitors who are taken seamlessly to a small open-air theater that hosts different shows. The square on the roof, which is comprised of a ramp that turns into a floor and hosts a public program, completes the exhibition space inside. The Germany Pavilion, together with Brazil Pavilion, chooses to transfer the experience of its use in public space to avoid being an object detached from the self-referential context while still generating a semi-public dimension that breaks with the dichotomy between main infrastructure (public open space) and pavilion (closed private space). Though this interpretation was suggested in the first draft of the urban concept of Expo 2015, it was not realized for various reasons. It has lost the idea of working on a large global garden to return to a traditional

岛屿、大海与食物馆 Islands, Sea and Food

国家保持着长期良好的关系。福斯特及其合伙人事务所建造了一个开放/封闭的空间，人们行走在七块弯弯曲曲的或封闭或开放的空间中，会想起沙漠景观和传统的沙漠式城市结构。福斯特所做的关于适用于沙漠性气候的可持续发展的智能城市模型研究与2015世博会的阿联酋馆设计很好地结合为一体。针对福斯特的设计作品，我们常常讨论与形式和所使用的设计语言有关的技术和工程的重要性，但是这家伦敦公司的专业知识现在能够克服人们对其在形式和技术工作方面的任何批评与挑剔。

从建设和管理使用后评价的角度来说，在形式主义往往凌驾于实效之上的当今时代，阿联酋馆无疑是一个非常伟大的建筑作品。

俄罗斯馆——通过俄罗斯在世博会上的表现，人们可以看到，无论是从文化上还是从历史上来说，俄罗斯并没有丧失建构主义的伟大遗产。展览也详细阐释了科学和农业之间的关系。悬挑入口的建筑表皮能反射入口处的一景一物，更加突出了展馆入口的宏伟壮观，而展馆不透明的部分则悬浮在透明的基座上，凸显了建筑的轻盈。同时，材料的选择和建筑技术的应用确保了施工进度。正如其他展馆，如智利馆，无论展馆空间多么复杂和精巧，建筑技术的进步都使其成为可能。这样说来，世博会就是一份真正的建筑技术给人们所带来的可能性的汇编，向人们一一展示了在建筑方法和建筑材料方面建筑技术到目前为止所带来的可能性。建筑方法和建筑材料越来越倾向于达到产品和

construction of urban space in which a street and a square representing the public dimension and the pavilions respectively serve the role of the container of activities.
Only Germany, Brazil, France and, to a lesser extent, China and Korea have developed in different forms this possible interaction that, in the context of the Expo theme, could be an important testing ground.
UAE Pavilion – The Arab Emirates have entrusted their pavilion to Norman Foster, one of the few superstars present at the Expo. Armed with a well-established relationship with the state of the Middle East, Foster + Partners built an open/closed space as it moves between seven perimeter and curved areas that recall desert landscapes and traditional urban structures. The research generated by Foster on sustainability applied to desert climates on models of Smart-cities find synthesis in the pavilion for Expo 2015. As often happens in the works of Foster, we meet to discuss the weight of the technical and engineering related to the forms and languages used, but the know-how of the London firm is now able to overcome any critical aspect in checking the formal and technical work.
This is always a great architecture product in an era when, often, the formalism prevails over effectiveness in terms of construction and management post-occupancy.
Russia Pavilion – The great legacy of constructivism is not lost culturally or historically to define the languages through which Russia presents itself at the Expo. The relationship with science is also explained in the exhibition that recounts the relationship between science and agriculture. The spectacular use of the cantilevered space is amplified by the building skin

干旱地区馆农业与营养展 Agriculture and Nutrition in the Arid Zones

生物地中海馆 Bio-Mediterraneum

香料馆 Spices

组装技术的智能混合,产品取自于大自然,组装技术部分在工厂里实现,部分在完成过程中实现。

西班牙馆——西班牙致力于建造一个双胞胎的形象,即一物可以复制其自身结构。整座展馆由两个细长的人字形屋顶结构组成,内有坡道通向展馆的封闭空间,像2015世博会上绝大部分展馆一样,西班牙馆也选择使用干性材料,如可部分在现场组装的胶合板。建筑材料和技术的选择必须考虑所要求的施工进度。

由于大部分的展馆寿命与世博会进行的时间是一致的,建筑材料的组装方法也同样决定了其拆卸方法,确保这些建筑材料在拆卸后几乎能够完全恢复使用。

万科馆——中国的经济实力反映在2015世博会中国场馆的数量上。中国共有三个展馆参展。本来应该有四个场馆,但是委托建造该展馆的公司中断了建造过程。

万科馆是丹尼尔·利贝斯金德的作品,也是整个世博会上最重要的作品之一。展馆的形状和红色的鳞状建筑表皮就像一只腾空的巨龙,完全不同于利贝斯金德通常所使用的建筑语言。

that reflects what happens at the entrance, while the opaque parts of the pavilion are suspended on a transparent base that amplifies the effect of lightness. At the same time, the choice of materials and construction techniques facilitated the quick construction of the building. As in other pavilions, such as the Chilean, the choices of space, however complex and sophisticated, are made possible by the evolution of building technologies. In this sense, the Expo is a real compendium of the possibilities offered to date in terms of construction systems and materials that are oriented more and more on an intelligent mix of products derived from natural and assembly techniques realized in part in factories and in part in work.

Spain Pavilion – Spain has been working on an iconic figure twinned, a solid that replicates itself in its structural part. The full portion, an iconic form with a pitched roof, is doubled and repeats only structural elements that support a ramp leading to the enclosed space of the pavilion. Like most of the exhibition buildings at Expo 2015, the Spain Pavilion is made with techniques preferring dry materials, such as laminated wood partially assembled on site. The construction choices are made necessary by the extreme speed at which buildings must be built.

Since the lives of most pavilions coincide with the duration of the event, the same assembly methods govern the removal of fabrics that can be recovered almost completely.

Vanke Pavilion – The economic power of China is reflected in the number of pavilions built for Expo 2015. There are three pavilions. There would have been four, but the company that commissioned them interrupted the construction process. The Vanke Pavilion is the work of Daniel Libeskind and is one of the most important works of the entire expo. Characterized by its shape and its skin in red shingles that refer to the body of a

谷类和薯类馆 Cereals and Tubers

大米馆 Rice

咖啡馆 Coffee

可可豆和巧克力馆 Cacao and Chocolate

水果和蔬菜馆 Fruits and Legumes

Copagri馆——由EMBT建筑师事务所设计的两座穹顶状建筑是一次令人好奇的实践，介于建设性实验和找寻自然形式之间，这也正代表着EMBT建筑师事务所现在所走的路径。两座穹顶状建筑是多边形的木质结构，一起构成了一个灵活且结构良好、光线通透的空间。这是一次灯光与邻近的万科馆和意大利馆互动的正式实验。

慢食馆——展馆名称来源于卡罗·佩特宁所发起的慢食运动，他也是米兰举行2015世博会的发起人之一。该馆由赫尔佐格&德梅隆设计完成，并继续在他们的研究中探讨对建筑类型的重新解释。项目所在地给设计师带来了许多实际限制。赫尔佐格&德梅隆通过嵌入微型城市来解决这个问题：建造三座完全用木材建造的坡屋顶建筑。三座木质结构建筑定义了一个三角形的共用空间。创造这样一小组空间——独立但又共用，赫尔佐格&德梅隆是为了体现世博会规划师们最初的倡议：世博会是一个大型城市花园，推动使用大型展览空间新途径。该场馆的成功运作部分应归功于这两位瑞士建筑师，他们的设计没有屈服于各种诱惑，而是致力于世博会最初的主题。

dragon, it has an identity far from the usual architectural language of Libeskind and is the closest pavilion at the Expo to the concept of folly, a singular object in a particular field.

Copagri Pavilion – The two domes of EMBT are a curious exercise suspended between constructive experimentation and a search for a natural form that identifies the present path of EMBT. Two domes made of a wooden structure polygonal create a flexible casing and are light-permeable. This is a formal experiment in which light interacts with neighboring Vanke Pavilion and Italy Pavilion.

Slow Food Pavilion – The pavilion takes its name from the Slow Food Movement started by Carlo Petrini, one of the initiators of the Expo 2015 in Milan. The work is done by Herzog & de Meuron, who continues in their research on the reinterpretation of the construction types. The site given to the architects presented some physical constraints. Herzog & de Meuron resolved them through a micro urban insertion: three buildings with a pitched roof, built totally in wood. The space of each building is defined by the reciprocal squares. In creating this small group – separate but open to public use – Herzog & de Meuron embodies the initial spirit of the planners of the Expo: to promote new ways of using such large exhibition spaces within the idea of a large urban garden. A successful operation is due in part to the Swiss architects, but they have the merit of not giving in to temptations and instead focusing their efforts on the theme of the original exposure.

Photographs by the courtesy of Expo Milano 2015 (©Daniele Mascolo - p.120, p.122, p.123, p.126, p.127, p.128, p.129 / ©Pietro Baroni - p.131, p.132, p.133), except as noted

生命之树

Marco Balich + Studio Giò Forma

生命之树是意大利馆的象征，集娱乐与整个世博园园区象征为一体。多亏了卡迪利特、额格里·布雷西业和皮里尔等赞助者的不懈努力，这个令人印象深刻的舞台设备才最终得以在短时间内与世人相见。它展示了扎根于过去和传统以在不同人群中开创充满文化、创新与富有成效的合作的重要性。

事实上，意大利馆不仅仅是一个建筑与空间组合而成的实体结构，它更是一个故事。这个故事简短精炼，定义了可以将意大利民族置于2015年米兰世博会核心位置的精神：一条以树为象征的发展之路。

若将大地比喻成滋生地的话，场馆变成了充满腐殖土的苗圃，大树深深地扎根于其中：经验、知识、文化、物质和非物质遗产等等，这些都是我们强大的力量和至关重要的能量。有了风和大气，枝繁叶茂的大树能够被这些肥料所滋养，并以参与、分享和慷慨的姿态将其种子传播到世界各处。

该结构由极富创意的设计师马尔科·巴里奇设计，其灵感萌芽于对意大利绝对代表性象征的探究，即代表了人类智慧的最为非凡的时期之一：文艺复兴时期。技术方面，"生命之树"建筑寓意从米开朗基罗设计的卡比托里欧广场上拔地而起，喷薄而出，高达35m，枝叶可伸展至42m左右。该建筑雄伟壮观，寓意丰富。它屹立在那里，直冲云霄，邀请游客举目注视并由衷赞叹周围的美景。精选的音乐，三维效果，光与影的出色运用，都确保会带给游客前所未有的情感体验。

Tree of Life

The Tree of Life is the symbol of the Italy Pavilion, a place for entertainment and a global icon. This impressive stage machinery was raised in record time thanks to the valuable work of its sponsors: Coldiretti, Orgoglio Brescia and Pirelli. It represents the importance of embedding roots into the past and traditions in order to open up to a future full of culture, innovation and fruitful collaboration among the various populations.

In fact, the Italy Pavilion is not just a physical structure made up of architecture and spaces, but it is also a narration, a simple and immediate story which defines the spirit that places our nation at the center of 2015 Expo Milano: a path which finds its gran finale in the symbol of the Tree. Following the metaphor of the breeding ground, the pavilion becomes the nursery, the humus of resources in which the

Tree embeds its roots into: experience, knowledge, culture, material and immaterial patrimony are our great "strengths" and our vital energy. The leafy branches of the Tree are nourished by this sap, which thanks to the wind and atmospheric agents, redistribute the seeds all over the world, in a symbolic gesture of participation, sharing and generosity.

The structure, designed by the creative director Marco Balich, buds from the search for a decidedly Italian symbol, representative of one of the most extraordinary periods of human genius: the Renaissance period. The architecture of the Tree,

©Luca Parisse (courtesy of 2015 Expo Milano)

which technically represents an "extrusion" from Michelangelo's floor in the Piazza del Campidoglio, reaches a height of 35 meters and its foliage widens out to about 42 meters. The architecture is imposing a crucial metaphor of allegory. It stands out against the sky and invites visitors to lift their gaze and admire the surrounding beauty. A specially chosen sound track, three-dimensional effects, extraordinary plays of light and projections, guarantee visitors an emotional experience like never before.

意大利馆
Nemesi & Partners

北立面 north elevation

南立面 south elevation

西立面 west elevation 0 10 20m

东立面 east elevation

Courtesy of Padiglione Italia

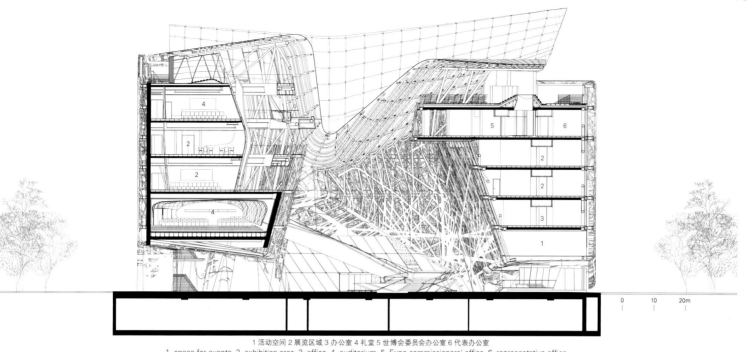

1 活动空间 2 展览区域 3 办公室 4 礼堂 5 世博会委员会办公室 6 代表办公室
1. space for events 2. exhibition area 3. office 4. auditorium 5. Expo commissioners' office 6. representative office
A-A' 剖面图 section A-A'

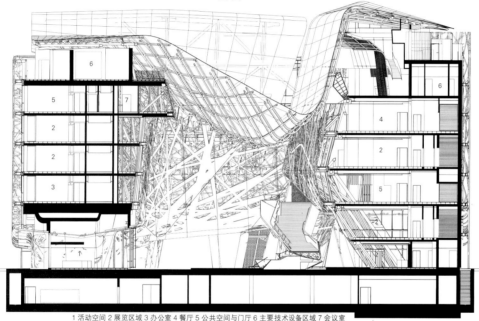

1 活动空间 2 展览区域 3 办公室 4 餐厅 5 公共空间与门厅 6 主要技术设备区域 7 会议室
1. space for events 2. exhibition area 3. office 4. restaurant 5. common space and foyer 6. main technical area 7. meeting room
B-B' 剖面图 section B-B'

意大利的苗圃

2015米兰世博会为意大利提供了一个绝佳的机会来复兴与发扬其在制造业、科学与技术等许多方面的长处。而这个使命的核心就是意大利馆。

苗圃意指培育项目与人才，为其提供肥沃的土壤，为其遮风挡雨，提供安息之所，赋予其新的能量，使其茁壮成长。大树象征着生命，象征着最原始状态的大自然，人类的一切活动都以此为中心而开展。

在2015米兰世博会上，有两点突出体现了意大利参会的特征：南北Cardo轴线与意大利宫殿。Cardo轴线是2015米兰世博会场地所有场馆组织安排的轴线之一，Cardo大街宽35m，长325m，很多展览与不同机构组织的活动都在此处举办。意大利宫殿面向阿克瓦广场的壮丽景色，是意大利国家和政府所在地的代表。

意大利馆由四个结构（代表四个街区）围绕一个大型中心广场组成，展示了真实的城市景象。中心广场是迎接游客的地方，也是公共空间的象征，整个展览从此处开始。这座树状的建筑扎根于地面，树干与上面的枝叶悬于空中。室内设有展览路线，游客一路可探索全部四层展览区域直至屋顶露台。到达屋顶露台之后，可通过另一条全新的不同的线路回到中心广场。

Italy Pavilion

The Nursery of Italy

The Universal Exposition of 2015 represents an excellent opportunity for Italy to revive and promote its many points of excellence in manufacturing, technology and science. And at the very center of this mission is the Italy Pavilion.

The nursery symbolizes the nurturing of projects and talents so that they can grow, providing them with fertile soil, offering them shelter and giving prominence to new energies. The tree is a symbol of life, of nature at its most primitive, a central icon around which all activities are arranged.

There are two focal points that characterize Italy's presence at Expo Milano 2015: the Cardo, one of the axes on which the Expo Milano 2015 area is organized, and Palazzo Italia. The Cardo avenue is 35 meters wide and 325 meters long, and hosts many exhibitions and institutional activities. The Palazzo Italia, facing the spectacular scenery of the Piazza d'Acqua, is the representative site of the Italian State and Government.

Its four blocks lay out real urban scenes that surround the large central square, which acts as a place of welcome and a symbol of community, the starting point of the exhibition. With its roots resting on the ground and branches and upper foliage lifted aloft, the building-tree offers an indoor exhibition route, a journey of discovery on all four levels of the exhibition area that leads right up to the rooftop terrace, and from there, back down, on a new and different path, to the central square.

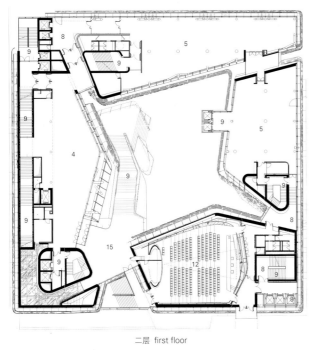

二层 first floor

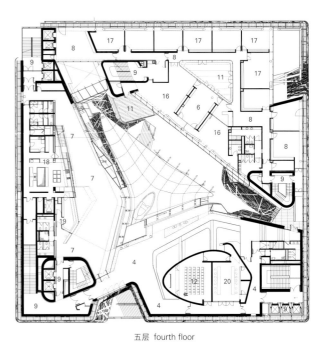

五层 fourth floor

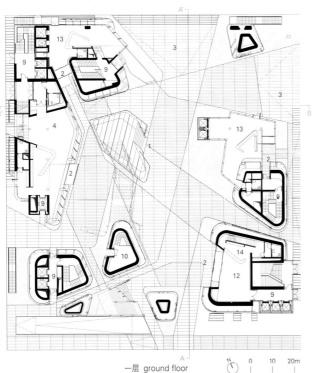

一层 ground floor

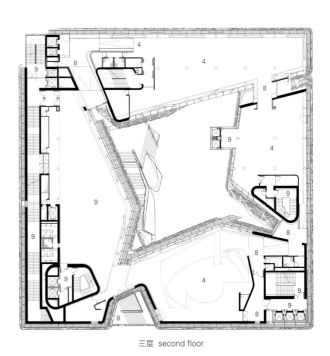

三层 second floor

1 主入口 2 入口 3 活动空间 4 展览区域 5 办公室 6 会议室 7 餐厅 8 公共空间与门厅 9 垂直连接空间 10 主要技术设备区域 11 露台 12 礼堂 13 办公室门厅 14 售票处 15 世博会书店 16 存款处 17 存款处门厅 18 世博会主楼梯 19 广场 20 礼堂 21 会议室 22 世博会委员会办公室 23 代表办公室 24 厨房 25 酒吧 26 代表团房间

1. main entrance 2. entrance 3. space for events 4. exhibition area 5. office 6. meeting rooms 7. restaurant 8. common space and foyer 9. vertical connection 10. main technical area 11. terrace 12. auditorium 13. office foyer 14. ticket 15. bookshop Expo 16. deposit 17. deposit foyer 18. Expo main stair 19. square 20. auditorium 21. meeting room 22. Expo commissioners' office 23. representative office 24. kitchen 25. bar 26. delegations room

面积：26,900m²
摄影师：
©Pietro Baroni (courtesy of 2015 Expo Milano)
-p.141(except as noted)
©Luigi Filetici(courtesy of the architect)-p.136, 138 bottom
Courtesy of Padiglione Italia-p.138 top

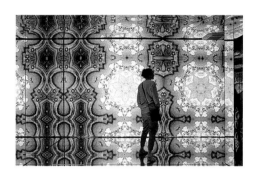

Courtesy of Padiglione Italia

意大利米兰——建筑师丹尼尔·利贝斯金德设计的四个树形结构,看起来像是闪闪发光的银色尖顶,构成了中心广场的框架。技术引领者西门子受到委托,负责制作这些名为"翅膀"的结构。每一个"翅膀"都由三个弯曲缠绕的叶片组成。这些叶子盘旋而上并向外伸展成树枝的样子,简直就是几何学的奇迹。这些结构每个都高达10m,重达14t,身带翅膀显得更为伸展。这些结构用声音以及与世博会主题"健康、能源、可持续发展与技术"有关的持续流动的脉动模型和充满想象力的画面,使得中心广场欣欣向荣,富有生气。"由于我一直着迷于将自然与科技结合起来,这些作品才得以出现。通过设计这些在主广场上翱翔的'翅膀',将公共空间变得充满生机,我希望能丰富游客的体验。这些'翅膀'同时起到指示方向的作用。"丹尼尔·利贝斯金德说道。这些结构表面的数字光带由10 000个独立LED灯组成。通过程序设计排列,可以营造出特效般大批椋鸟起飞的壮观景象。

为了创作音乐背景(音景),作曲家在作曲时通过将自然界的声音与机器和城市生活的声音分层,将自然与科技很好地融为一体,在两者之间架起了一座桥梁。这些巨大的结构,最初只是做成3D模型,后来用1000块铝板建造而成。每块铝板都在德国——焊接。

The Wings

Milan, Italy – Like shimmering, silver finials, a quartet of tree-like sculptures created by the architect Daniel Libeskind will frame the central square. Entitled the Wings, the sculptures were commissioned by the technology leader Siemens. Marvels of geometry, each is composed of three sinuous, intertwined blades that spiral up and outward like branches of a tree. Measuring ten meters high and weighing 14 tons each, crowned by wings that span just as far, the sculptures will animate the piazza with sound and a constant flow of pulsating patterns and imagery related to the Expo's themes: health, energy, sustainability and technology. "These works grew out of my ongoing fascination with the intersection of nature and technology. I wanted to contribute to the visitor experience by enlivening the public space, by creating these sculptural wings that soar over the main piazza and also serve as directional markers," said Daniel Libeskind. The digital light pattern on the surface of the sculptures is made up of 10,000 individual LED units that are programmed with a sequence that recreates the spectacular acrobatic masses of starlings taking to flight.

To create the soundscape, the composition explores the bridge between nature and technology by layering sounds found in nature with those of machinery and urban life. The forms were 3D modeled and then translated to 1,000 aluminum plates that were individually welded together in Germany to create the massive sculptures.

翅膀
Studio Libeskind

高度:10m / 材料:structure_monocoque aluminum, facade_brushed aluminum fitted with LED technology / 摄影师:©Hufton+Crow(courtesy of the architect)

万科馆的设计包含源自中国饮食文化的三个理念：食堂——传统的中国餐厅；风景——基本的生活元素；龙——暗喻与农作和食物有关。所有这些理念都与世博会场馆的展览、建筑的结构和举办的项目融为一体。

万科馆坐落于阿瑞娜湖的东南部，这个800m²的场馆似巨龙般从东方升起，形成一副动态而垂直的景观。该设计的特别之处在于场馆内外贯穿着蜿蜒迂回的几何图案。楼梯十分宏伟，暖灰色水泥的表层中混杂雕刻着红色的蜿蜒曲线，游客顺着楼梯即可到达顶层。从顶层郁郁葱葱的花园观景台望出去，可以看到阿瑞娜湖和邻近的意大利馆的迷人景色。

场馆表面铺设了4000余块红色镀金属瓷砖。这些几何图案的陶瓷嵌板不仅创造了和龙的皮肤一样极具表现力的图案，而且还具备高度可持续自净和空气净化的性能。其金属着色的三维表层会随着光线与视角的改变而改变。有时候看起来是深红色，有时闪耀着金色，甚至从某些特定的角度看会变成绝妙的白色。

安装镀金属瓷砖使用的是最先进的覆层支撑系统，造型精确，富有节奏感。否则的话，场馆会看起来软塌，扭转弯曲也会变得僵硬。

Vanke Pavilion

The concept for the Vanke Pavilion incorporates three ideas drawn from Chinese culture related to food: the shi-tang, a traditional Chinese dining hall; the landscape, the fundamental element to life; and the dragon, which is metaphorically related to farming and sustenance. All concepts are incorporated in the pavilion's exhibition, architecture and program. Situated on the southeast edge of the Lake Arena, the 800-square-meter pavilion appears to rise from the east, forming a dynamic, vertical landscape. The design features a sinuous geometrical pattern that flows between inside and outside. A grand staircase, clad in warm grey concrete, carves through the red serpentine form and guides visitors to the upper level. A roof-top observation deck with a planted garden will provide stunning views of the lake and near-by Italy Pavilion.

The pavilion is clad in more than 4,000 red metalized tiles. The geometric ceramic panels not only create an expressive pattern that is evocative of a dragon-like skin, but also possess highly sustainable self-cleaning and air purification properties. The three-dimensional surface is coated with a metallic coloration that changes as light and viewpoints shift. At times it will appear as deep crimson, then a dazzling gold, and even, at certain angles, a brilliant white.

The tiles are installed with a state-of-the-art cladding support system that gives a rhythmic pattern and mathematical form to an otherwise supple, torquing shape.

万科馆

Studio Libeskind

有效楼层面积：1,360m² / 材料：structure_steel, facade_red metalized porcelain tiles, floors_porcelain tiles / 摄影师：©Hufton+Crow(courtesy of the architect)

为Copagri设计的场馆叫作"爱IT",坐落于阿瑞娜湖的东南部。阿瑞娜湖呈圆形,位于2015米兰世博会总体规划设计的主轴线之一"Cardo"的一段。

这个建筑项目是两个直径相同而高度不同的穹顶的合二为一。两个穹顶一起创造了灵活而结构优良的空间,空间可以进一步细分以满足不同的内部空间需要。

该结构由巨大的三维网格构成。三维网格渐渐变形为交织的枝干,一直延伸到敞口穹顶的最高处。结构元素变成了建筑元素,将场馆内部与外墙均设计成树枝。网格之间的地方使得空气和光线更易透过,同时也保持了内外的视线连续性。

双穹顶是预制结构,由云杉木胶合板及镀锌钢连接件构建而成。穹顶上部区域被称为"帽子",内置为建筑自身和为阿瑞娜湖提供必要服务所需的设施,诸如灯光、扬声器和天线等。

建筑内部空间设计是要营造意大利非常典型的"当地市场",每一个生产者都可以在此展示他们的产品,讲述他们的故事,介绍他们的公司。

Copagri Pavilion

The pavilion designed for Copagri called "Love IT" is located in the south-east part of the Lake Arena – a large circular pool of water at the end of the Cardo, one of the main axes of the 2015 Expo Milano masterplan.

The project stems from the combination of two single domes of equal diameter, but different heights. The two domes to-

Copagri馆
EMBT

©Roland Halbe (courtesy of the architect)

gether create a flexible and well-structured space that could be further subdivided in order to fit different needs of the internal spaces.

The structure is composed of a big three-dimensional grid that gradually transforms in woven branches, as it grows to the open top of the dome. The structural elements become architectural ones, designing both the internal and the external facades as tree branches; the empty spots of the grid enhance air and light permeability, together with visual continuity between outside and inside.

The double dome is a prefabricated structure, made of spruce glulam with zinc-coated steel joints. The upper part of the domes, called "the hat", can host the required building services both for the building itself and for the Lake Arena, consisting of lights, loudspeakers and antennas.

The space inside is conceived to host a typical Italian "local market", where each producer exhibits the products, tells their story, and introduces their company.

A-A' 剖面图 section A-A'

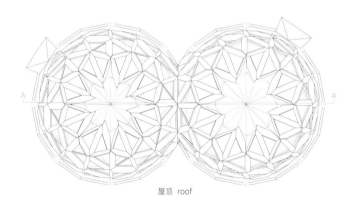

屋顶 roof

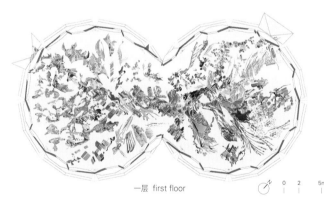

一层 first floor

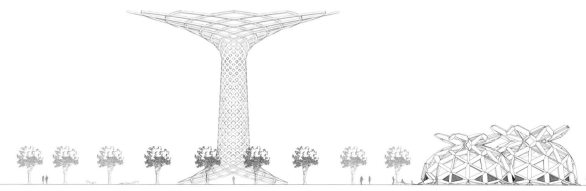

南立面 south elevation

奥地利馆
team.breathe.austria

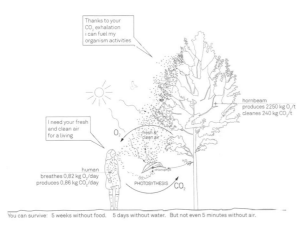

概念——空气即生命
concept _ air is life

呼吸.奥地利

奥地利以文化之乡享誉世界，且生活水平较高。诸如风景和气候等的新元素推动了对利用可持续资源的认知，也大大提高了奥地利的正面形象。在这儿，运用可持续发展技术已经可以生产大量的食物，而且消费者能够在超市购买到的有机产品范围也比欧洲任何其他地方大得多。鼓励生产者、经销商和消费者之间建立牢固的联系对奥地利食品可持续发展之一主题思路至关重要。

空气对于食物和人类的健康至关重要，它还是生态平衡的指标。以空气和奥地利出彩之处开篇，游客们既可以在奥地利馆感受到这个国家最广为人知的一面，同时也可以探索出新的方面。一场体验之旅可以充分感受来自格拉茨市的Terrain团队所创建的一个将建筑、自然、文化与研究融为一体的空间。场馆内建有一个小型奥地利森林，不用过滤器或调节器即可每小时提供62.5kg新鲜健康的氧气并吸收CO_2，在理想的气候中足够1800人呼吸使用。这个绿肺倡导世界变得更加干净，为城市实践提供了确保拥有更高生活质量的模型，并努力证明在全球绿化面积都在衰减的时候重新造林大有裨益。

Austria Pavilion

Breathe. Austria

Austria is world-renowned as a country of culture, with a high standard of living. Its positive image is enriched by new elements, such as the landscape and climate, playing a key role in driving awareness around the use of sustainable resources. A large quantity of food is already produced with sustainable techniques and the range of organic products available to consumers in supermarkets is higher than anywhere else in Europe. Encouraging strong connections between producers, sellers and consumers is central to Austria's thematic itinerary of food sustainability.

Air is an essential component to the health of food and humans and it is an indicator of ecological balance. Starting with the element of air, of notable excellence in Austria, the pavilion will give visitors the opportunity to connect with the most well-known aspects of the country, while discovering new facets. Team Terrain of Graz has created a space where architecture, nature, culture and research come together in a unique experiential journey. The pavilion creates a small scale Austrian forest that provides 62.5kg of fresh oxygen every hour, without filters or conditioners, which is enough for 1,800 people in an ideal climate, providing wellness and absorbing CO_2. It is a green lung that induces the desire for a cleaner world, offering a model for urban practices that can ensure a higher quality of life and demonstrating the benefits of a reforestation policy against the global decline of green areas.

摄影师：
©Daniele Mascolo (courtesy of the architect) - p.147, p.150
©Marc Lins (courtesy of the architect) - p.146, p.148, p.149
©Pietro Baroni (courtesy of 2015 Expo Milano) - p.151

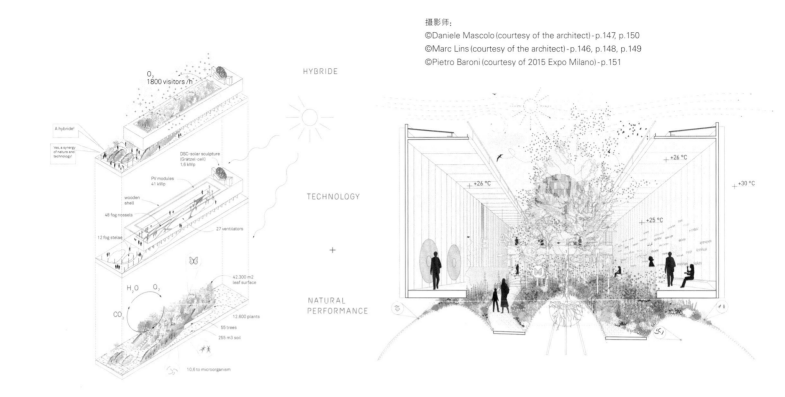

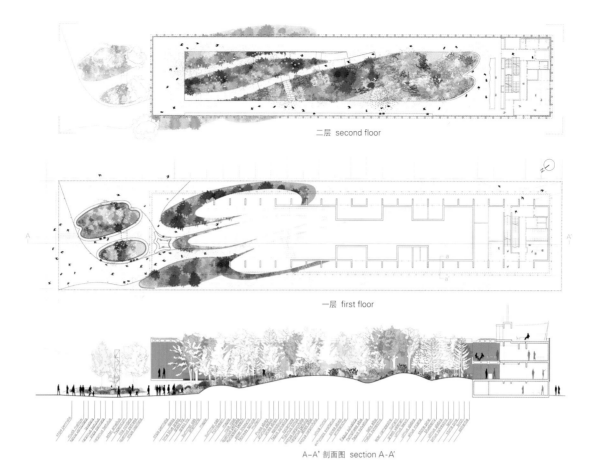

二层 second floor

一层 first floor

A-A' 剖面图 section A-A'

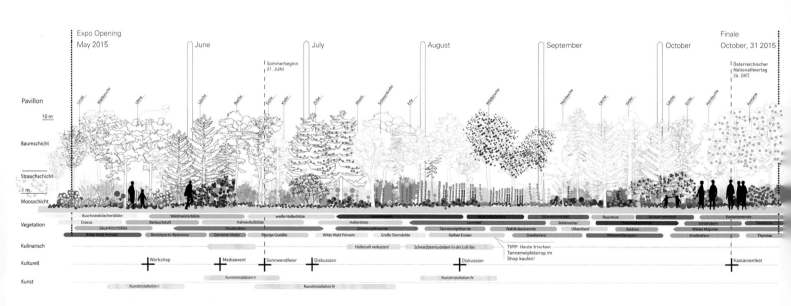

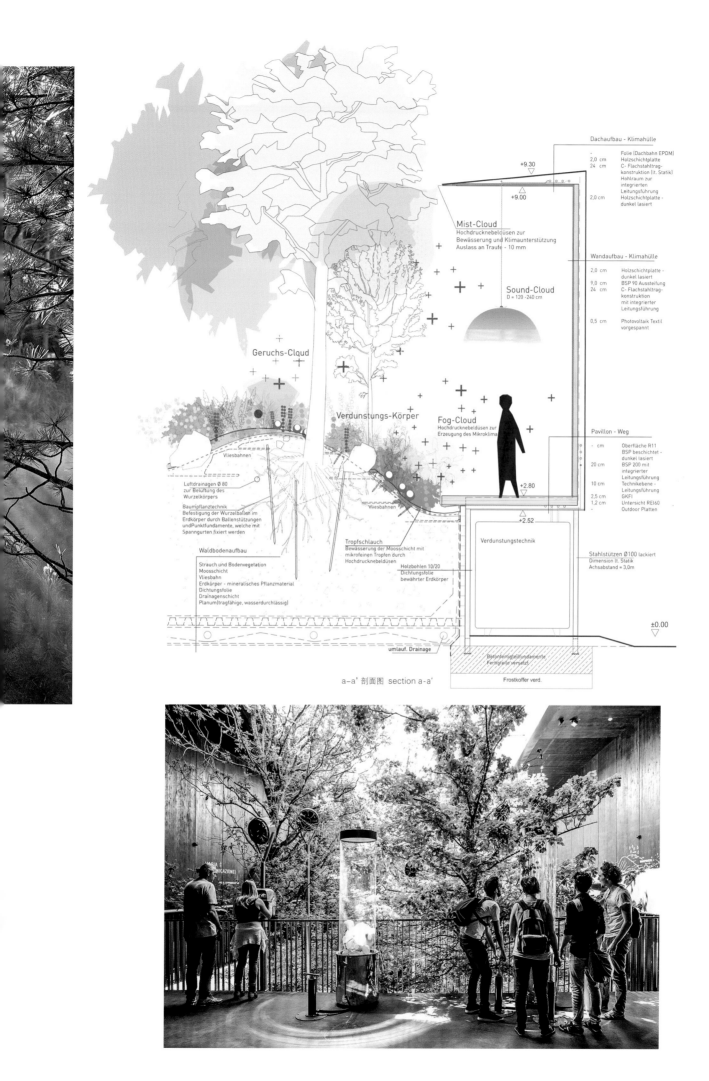

巴西馆

Studio Arthur Casas + Atelier Marko Brajovic

喂饱全世界

巴西是世界上最大的粮食生产国之一。巴西农工业活动已为大家熟知，但是其技术能力大家所知甚少。

巴西提出了一些与农业技术能力和扩大粮食生产及相关产品出口能力有关的解决方案，以此来诠释2015米兰世博会的主题："给养地球，生命的能源"。另一方面，巴西馆展示了在不破坏生物多样性的前提下满足社会需求的能力，这对在整个地球范围内实现生态平衡非常重要。

巴西此次参加世博会有三个关键点：技术——着重强调其技术的发展以及作为潜在的技术合作伙伴的可靠性、创新、高效；文化——通过一系列的典型产品突出其文化的多元化；社会——确保人人享受全球性健康食物。

基于"喂饱全世界"的主题，巴西馆使用了网络这一暗喻，在灵活性、流动性和分散性等方面，展示出不同主题之间的联系，完整而统一，使巴西成为粮食生产方面的全球领导者。

在4133m²的展览空间内，巴西让2015米兰世博会的参观者看到了为满足全世界的粮食需求，用真正可持续的先进技术来探索和实施提高粮食产量和增加食物多样性的各种可能性。

Brazil Pavilion

Feeding the World with Solutions

Brazil is one of the greatest producers in the whole world. Whilst its agro-industrial activity is widely known, its technological capacity is less so.

Brazil is interpreting the theme of Expo Milano 2015, "Feeding the Planet: Energy for Life", by providing a number of solutions connected to its technological capacity within agriculture contexts and its ability to extend the production of food and its related exports. Another aspect is its capacity for satisfying social demand without damaging biodiversity, which is a fundamental resource for achieving balance throughout the entire planet.

Brazil's participation at the Expo is based on three key points: technological, underscoring its development and reliability

west elevation

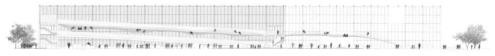

A-A' 剖面图 section A-A'

南立面 south elevation

一层 ground floor

as a potential technology partner that is both innovative and efficient; cultural, highlighting its plurality through a range of typical products; and social, guaranteeing global, healthy food that is accessible to all.

Based on its theme "Feeding the world with solutions", the Brazil Pavilion uses the metaphor of the network, in terms of flexibility, fluidity and decentralization, showing the relationships and integration of different topics that combined, make it the global leader in food production.

In an exhibition space of 4,133m², Brazil offers Expo Milano 2015 visitors a view of all the possibilities being explored and implemented to increase and diversify food production, satisfying food demands around the world and using advanced technologies in a way that is truly sustainable.

用地面积：4,133m² / 建筑面积：3,674m² / 摄影师：©Raphael Azevedo França (courtesy of Hill+Knowlton Strategies) - p.153, p.155 left-first, p.155 right-top
©Pietro Baroni (courtesy of 2015 Expo Milano) - p.152, p.154, p.155 left-second, third, fourth

英国馆

Wolfgang Buttress

生于不列颠及北爱尔兰

英国馆以"生于不列颠及北爱尔兰"为主题亮相2015米兰世博会。目前我们面临一个非常迫切的问题，即如何养活到2050年即将突破90亿的世界人口，并将人口数量维持在预期增长范围内。英国馆旨在提升全球对于该问题的认知，并努力提供创新的解决方法。

英国馆利用蜜蜂的旅程来强调授粉对于全球食物链和生态系统的重要性。这个复杂的建筑环境微缩表现了英国在解决全球面临的粮食挑战方面所起的作用。运用暗喻的手法，展馆描述了思想、技能和知识的交流对于人类活动的重要性。而作为全球联系枢纽的英国，便是进行这种交流（异花授粉）的理想之地。英国总人口不到全球人口的百分之一，然而英国的发明家和企业家却为当今世界做出了巨大贡献。

不列颠就是一个大蜂巢，在这里人们忙碌着，寻找解决当今所面临的一些巨大挑战的方法，并将这些解决方法分享给全世界。游客到达17m高的铝质球状造型的蜂巢内，即可深度游览体验拥有独特英国本土物种的果园和野花草地。这个蜂巢由180 000多个部件构成，重约30t。游客会被声音和闪烁的灯光所包围，看起来像是一个真正的位于英国的蜂巢在运作。诺丁汉特伦特大学的马丁·本次斯博士突破常规，研发了这个场馆所采用的技术。它可以监控蜂巢的健康状况，帮助解决维持蜜蜂数量这一挑战——这对于授粉与喂养这个星球十分重要。在为全世界提供90%的粮食来源的100种农作物中，有70多种都只靠蜜蜂授粉。

英国的科学、创新与创造力融为一体，交织成英国馆这座特有的建筑，与2015米兰世博会的主题直接呼应。英国馆很好地证明了想法与创新是如何在不列颠生长的。

A-A' 剖面图 section A-A'

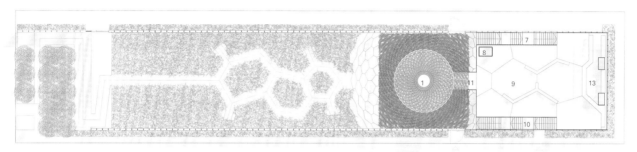

屋顶 roof

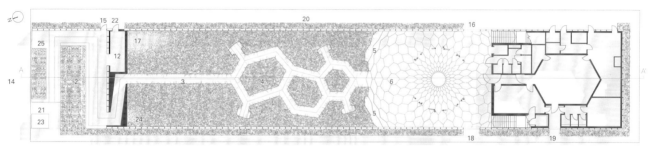

二层 first floor

1 蜂巢雕塑 2 果园/展馆入口 3 野花草地步道 4 野花草地 5 阶梯座椅 6 活动空间（空间下方）7 通往屋顶露台的阶梯通道 8 电梯 9 屋顶露台 10 通往一层的阶梯通道 11 蜂巢入口通道（屋顶层）12 售票处 13 酒吧/商店屋顶 14 Decumanus大街 15 残障游客入口 16 紧急出口 17 灌溉水池（草地下方）上方的入口 18 展馆出口 19 休息室与会议室VIP入口 20 限制入口区 21 果园面向Decumanus大街的入口 22 售票处员工入口 23 售票处电话亭 24 配电站（草地下方）25 配电站上方座椅/入口

1. bee hive sculptural piece 2. orchard/pavilion entrance 3. wildflower meadow walk 4. wildflower meadow 5. terraced seating 6. event space (beneath space) 7. step access up to roof terrace 8. lift 9. roof terrace 10. step access down to ground floor 11. bee hive access walkway (at roof level) 12. ticket office 13. roof to bar/shop 14. Decumanus 15. disabled visitor access 16. emergency exit 17. access cover to irrigation tank (beneath meadow) 18. pavilion exit 19. VIP entrance to lounge and conference facilities 20. restricted access area 21. entrance into orchard from Decumanus 22. staff access to ticket office 23. ticketing kiosk 24. transformer room (beneath meadow) 25. seating/access cover over sub-station

UK Pavilion

Grown in Britain and Northern Ireland

The UK's presence at 2015 Expo Milano, under the theme "Grown in Britain and Northern Ireland", aims to raise global awareness of and provide innovative solutions to one of the most pressing challenges of our time – how to feed and sustain an expected rise in the world's population to 9 billion by 2050.

The UK Pavilion follows the journey of the honey bee to highlight the role of pollination in the global food chain and ecosystem. This complex environment encapsulates some of the qualities the UK brings to the global food challenge. It offers a metaphor for describing how the exchange of ideas, skills and knowledge is an essential part of human activity and the UK, as a gateway of global connections, is the ideal place for such cross-pollination to happen. The UK accounts for less than one percent of the global population, but its innovators and entrepreneurs have helped shape the modern world. Britain is a hive of activity where solutions to some of today's greatest challenges are developed and shared with the world. Visitors will take part in an immersive experience through a British orchard and wildflower meadow with typical native plant species, moving onto a 17-meter-high stylised bee hive in the form of an aluminum sphere. The hive is made up of over 180,000 components and weighs 30 tonnes. Visitors will be surrounded by sounds and lights flickering in response to movements within a real bee hive based in the UK. This ground breaking UK technology, developed by Dr. Martin Bencisik at Nottingham Trent University, monitors the health of bee hives and helps to solve the challenge of sustaining bee populations – vital to pollination and feeding the planet. Of the 100 crop species that provide 90 percent of the world's food, over 70 are pollinated by bees alone.

Woven into the UK Pavilion's very structure is a demonstration of how science, innovation and creativity combine to directly address the theme of Expo Milano 2015. The UK Pavilion is testament to how great ideas and innovations are grown in Britain.

摄影师: Courtesy of the UKTI - p.158[first, second, fourth]
©Hufton+Crow (courtesy of the architect) - p.156, p.157, p.158[third]

韩国饮食，面向未来的饮食：人如其食

每个人都有自己喜爱的美食。由于文化、环境和喜好不同，故有无以计数的不同类型的食物，难以分门别类地区分它们。现在，我们希望问游客一个问题："你会怎样吃？你会吃什么？你会坚持吃多久？"问这个问题的原因是我们每天吃的食物维系了我们的身体，推动着我们的生活，形成了属于我们自己的文化。

作为韩国馆主题的寒食节（韩国料理），有着悠久的文化传统，其根本在于与自然和谐相处。韩国馆将向人们呈现切实可行的、可持续发展的替代食物，子子孙孙都可以享用不尽。

韩国馆将提出人类当今正在面临的各种各样的与食物有关的问题的解决之道。讨论这些问题将有利于提高认识，为切实地解决问题提供一个方向。韩国馆的场馆建筑以"月亮罐"为主题，这种传统陶器的形状就像满月。流线型的曲线、精心设计的装饰点缀，让人感到整栋建筑的简洁明快，与周围环境和谐而平衡，看起来就好像漂浮在周围环境中。

米兰世博会园区的整体规划是受Decumanus大街这一古罗马几何结构的启发，而韩国馆形状的设计同样带有韩国的传统，有机生态。通过这种方式，韩国馆向人们传递了东方和西方和谐融洽的信号。

Korea Pavilion

HanSik, Food for the Future: You are What You Eat

Everyone has a type of food that he or she enjoys. The countless different types of food based on one's culture, environment, and preference make it hard to even distinguish among them. Now, we wish to ask a question to our visitors: "How will you eat? What will you eat? How long can it be sustained?" The reason for this question is that the food we eat everyday sustains our bodies, propels our lives and forms the cultures

韩国馆
Archiban

we belong to.

With HanSik (Korean food), as the theme of the Korea Pavilion, the longstanding traditions of Korean food, which has its base rooted deeply in harmony with nature, will present a feasible and sustainable food alternative that can be used for all future generations.

The Korea Pavilion will address various food-related issues that mankind is currently facing. Discussing these issues will help increase awareness and offer a direction to a reliable solution. The pavilion will be constructed with the architectural theme of the "Moon Jar", a traditional pottery in the shape of the full moon. The curvature and subtle accents lend to the overall feeling of simplicity and harmonious balance with the surrounding environment, and was made to appear as if it were floating within its surrounding environment.

While the Milano Expo site has been planned with a Decumanus – inspired Roman geometric structure, the Korea Pavilion has been designed with an organic traditional Korean shape. In this way it conveys a message of harmony between East and West.

用地面积：3,880m² / 有效楼层面积：3,990m²
摄影师：©Marco Atzori-p.160, p.161, p.162^{top}, p.162$^{bottom-right}$
©Pietro Baroni (courtesy of Expo Milano 2015)-p.162$^{middle-left}$, p.162$^{middle-right}$, p.162$^{bottom-left}$

东立面 east elevation

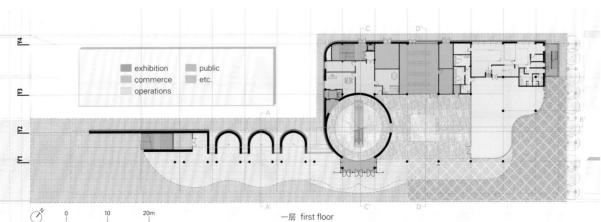

一层 first floor

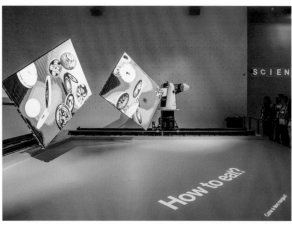

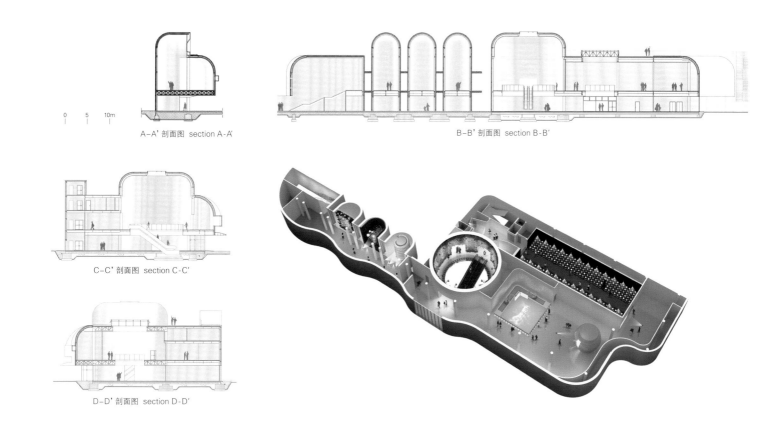

A-A' 剖面图 section A-A'
B-B' 剖面图 section B-B'
C-C' 剖面图 section C-C'
D-D' 剖面图 section D-D'

©Daniele Mascolo (courtesy of Expo Milano 2015)

法国馆
X-TU Architects

生产食物和提供食物的不同方式

我们现在和将来怎样养活整个世界？我们怎样才能确保长期为人类提供既充足又优质、健康的食物？

法国馆通过四根柱子来表达主题：通过法国富有成效的基础设施潜能为全球粮食生产做出贡献；开发新的食品类型来应对人们对更好产品的需求；通过技能和技术转让政策来提高发展中国家自给自足的能力；在所有领域，无论是健康方面还是营养或烹饪方面，确保数量和质量并重。

法国馆主要由胶合板建造，占地面积为3592m²，其灵感源于法国有顶棚遮挡的市场，是法国饮食文化的象征。法国许多城市都有传统市场，这显然代表了2015米兰世博会总的主题，强调食品安全、食品来源和食品质量。

考虑到展馆只是临时建筑，所以设计师选用了轻质的木结构。展馆由法国Simonin建筑公司负责建造，世博会结束后可以拆除并重新利用，特别关注减少能源消耗、废物回收利用及净化工艺。

France Pavilion

Different Ways of Producing and Providing Food

How can we feed the world, today and tomorrow? How can we ensure adequate food for mankind that is of good quality and healthy in the long term?

Its communication is based on four pillars: contributing to global food production, through the potential of France's productive infrastructure; developing new food models, to address the need for better production; improving self-sufficiency in developing countries, with a policy of skills and technology transfer; and aligning quantity with quality in all areas, be they health-based, nutritional or culinary.

市场发展 market evolution

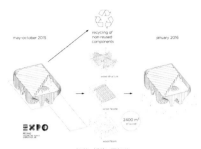

组装+拆除+再组装
assembly + disassembly + reassembly

建筑形式的形成
generation of the form

木结构
wood structure

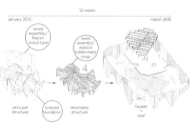

结构体系
structural system

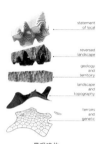

景观建构
landscaped building

用地面积：3,500m² / 有效楼层面积：3,532m² / 造价：EUR 14 million
摄影师：©Andrea Bosio (courtesy of the architect) - p.178, p.180, p.181^{top} ©Daniele Mascolo (courtesy of 2015 Expo Milano) - p.179, p.181^{bottom}

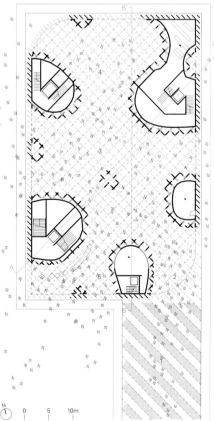

1 花园 2 入口 3 展览市场 4 临时展馆 5 商店
1. garden 2. entrance 3. exhibition market 4. temporary exhibition 5. shop
一层 ground floor

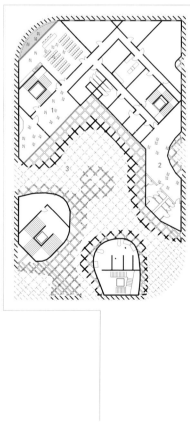

1 VIP室 2 办公室 3 展览区上空空间
1. VIP room 2. office 3. void on exhibition area
二层 first floor

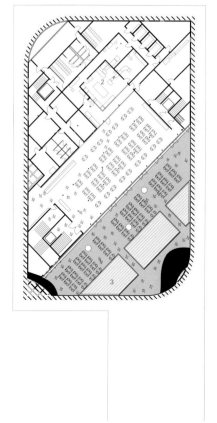

1 餐厅 2 厨房 3 露台
1. restaurant 2. kitchen 3. terrace
三层 second floor

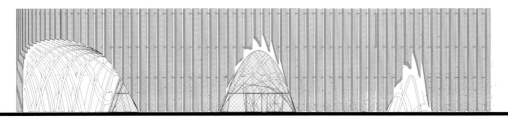
东立面 east elevation

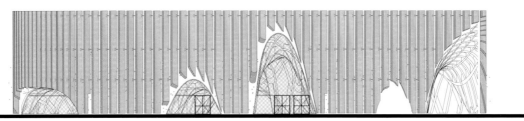
西立面 west elevation

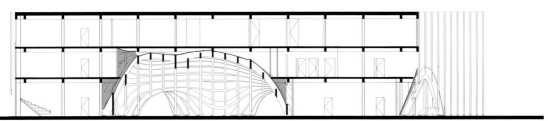
A-A' 剖面图 section A-A'

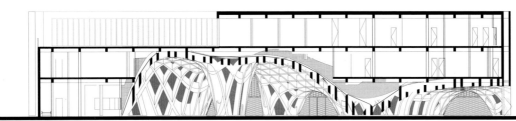
B-B' 剖面图 section B-B'

The building, made mainly of laminated wood, spreading over a 3,592-square-meter space, is inspired by the covered market, a symbol of French food culture. The pavilion takes its inspiration from the fact that traditional markets are located in many cities of France, which clearly represent the overall theme of Expo Milano 2015, with an emphasis on food security, access to food and the element of food quality.

Given that the pavilion is only temporary, a lightweight construction with a wooden structure was chosen, produced by the French company Simonin, and can be dismantled and reused at the end of the exposition. Particular attention has been given to reducing energy consumption, waste recycling and purification processes.

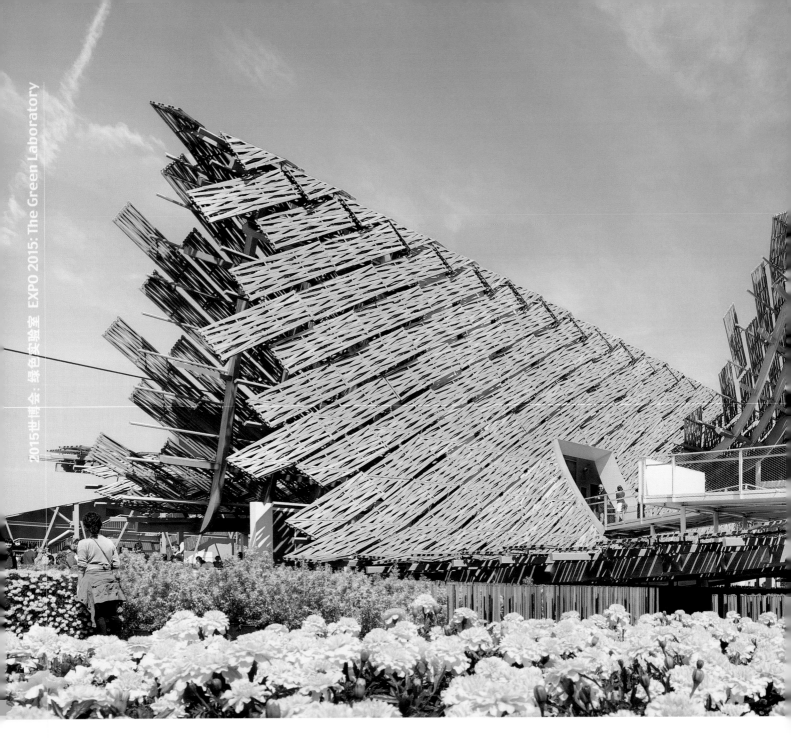

中国馆
Tsinghua University + Studio Link-Arc

希望的田野,生命的源泉

中国馆的主题体现了感恩、尊重和合作,这也体现了中国人民的特点:"田野"伊始就哺育着人类,"希望"食物在将来成为每个人"生命的源泉"。

农业、食物、环境以及可持续发展是中国参加2015米兰世博会的主题,其目的是唤起人们对中国"天人合一"这一古老哲学的记忆,形象展示中国在农业领域的文化传统和取得的进步,展示中国为了提供充足、健康、优质的食品在资源利用方面所做出的巨大贡献。

中国馆的设计理念意在追求人与环境、人与自然之间的平衡。正如农民会细心呵护他的土地,人类必须关爱地球。

占地4590m²的中国馆主要展现了三大主题。"大自然的恩赐"板块根据中国农历和土壤的五种颜色来说明农作物的生长过程。"生命的源泉"板块展现了食品的制作过程,包括豆腐以及其他食品的制作流程、中国著名的八大菜系以及茶文化。"科技与未来"板块通过图表形式展示了科技的进步,包括袁隆平教授的杂交水稻、循环农业以及物联网跟踪技术。

这个项目是由清华大学和北京清尚环艺建筑设计院有限公司共同设计的。自然天际线与城市天际线相互交融;各种产品、水稻和小麦都纳入展览空间之内,地板和其他元素也与中国传统建筑特色相呼应。不同展区分别被叫作"天""地""人""合"。

摄影师：
©Hufton+Crow (courtesy of the architect) - p.171
©Sergio Grazia (courtesy of the architect) - p.168

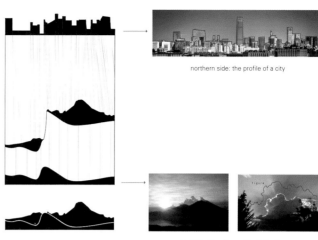

northern side: the profile of a city

southern side: the profile of a landscape

屋顶板材几何形状合理化设计
roof panel geometry rationalization

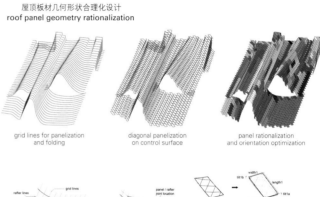

grid lines for panelization and folding

diagonal panelization on control surface

panel rationalization and orientation optimization

segmentation of curved control surface / diagonal panelization with folding

rectangular panel shape / reasonable folding / unitized panel size

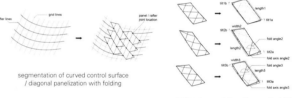

屋顶板材类型展示
roof panel type counts

China Pavilion

Land of Hope, Food for Life

China's theme captures an attitude of thankfulness, respect and cooperation that characterizes its people: the land has fed man from its beginnings and hope suggests the prospect of future where food can offer life to everyone.
Agriculture, food, environment and sustainable development are the focal points of China's participation in 2015 Expo Milano. Its aim is to recall the tenet of Chinese philosophy that

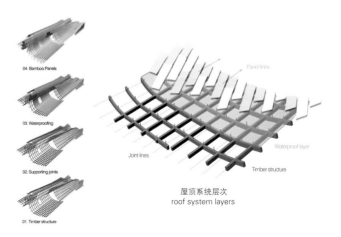

屋顶系统层次
roof system layers

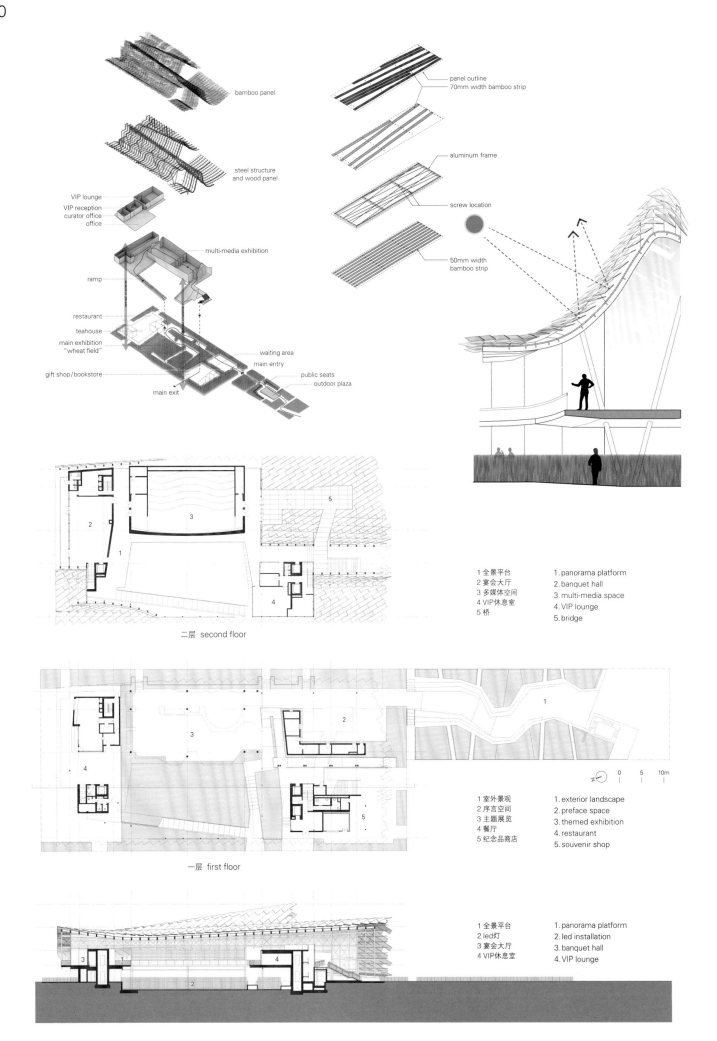

"Man is part of nature", and to illustrate its cultural traditions and progress in the areas of agriculture, showcasing the great strides made in the use of resources for providing a sufficient supply of good and healthy food.

The underlying theme is the pursuit of balance between mankind and environment, between humanity and nature. Just as the farmer looks after and protects the earth, so must people care for the planet.

The exhibition area of 4,590 square meters unfolds around three themes. "The Gift of Nature" illustrates crop processes according to the Chinese lunisolar calendar and the five colors of the soil. "Food for Life" shows the production path of foods including tofu and other dishes, China's famous eight schools of cuisine, and its tea culture. "Technology and the Future" charts the progress of science, including the hybrid rice of Professor Yuan Longping, recycling in agriculture and techniques for tracking the Internet of things.

The project was developed by a consortium created by Tsinghua University and by the Beijing Qingshang Environmental & Architectural Design Institute. The forms of the natural landscape on the one side are transfused and combined with those of a city skyline on the other. Products, rice, wheat, are located in the spaces, floors and other elements that echo traditional Chinese architecture. The various exhibition areas are called "Heaven", "Earth", "Man" and "Harmony".

俄罗斯馆 _ Speech Tchoban & Kuznetsov

栽培世界，哺育未来

与"给养地球，生命的能源"主题呼应，2015米兰世博会上俄罗斯馆的设计理念反映了俄罗斯为全球食品安全已经做出的巨大贡献并将继续通过利用本国的科技和农业传统为全球食品安全做出贡献。另外，展馆也展示了俄罗斯作为世界上最大国家为世界生产更多粮食的巨大潜力。到2050年，世界人口预计将超过90亿，全球粮食产量需要增加60%来养活如此庞大的人口。因此，20国集团国家，尤其是俄罗斯，已经把食品安全作为优先考虑的问题，并且俄罗斯的有利地位也使其完全有能力解决这一问题。展馆讲述了几位在国际上享有盛誉的俄罗斯科学家的故事，他们为农业和食品安全的发展做出了贡献。展馆也展示了俄罗斯丰富的自然资源以及非富多样的烹饪传统。这个展馆突出强调了俄罗斯在食品安全和食物供给等重要方面正在发挥并将继续发挥的重要作用；突出强调了俄罗斯正怎样为本国人民和世界人民提供粮食。俄罗斯馆占地超过4000m²，整个展馆大气且富有活力，以一种气势恢宏的姿态伸向天空，集独特的工程解决方案和绿色技术于一体。简洁而令人难忘的外立面是由可循环利用、生态环保的材料制成的。体积庞大的小山成为自然景观元素，四周由天然有机木箍围挡，形状就像帆船的"鼻子"非常优雅。这只帆船被喻为诺亚方舟。

>> Russia Pavilion

Growing for the World, Cultivating for the Future

Embracing the theme "Feeding the Planet. Energy for Life", the concept of the Russia Pavilion at 2015 Expo Milano reflects the great contribution to global food security that Russia has made and will continue to make through its scientific and agricultural heritage – and the enormous potential of the world's largest country to produce food for the world. With global food production needing to increase by 60 percent to feed a world population that is expected to exceed nine billion by 2050, the G20 group of nations – and Russia in particular – have identified food security as a priority issue, and one for which Russia is well placed to address. The pavilion tells the stories of several internationally recognized Russian scientists, whose works have contributed to the development of agriculture and food security. It also demonstrates the wealth of Russia's natural resources and its rich and diverse culinary traditions. The pavilion highlights the role that Russia is playing, and can continue to play, in the vital aspects of food security and food supply, and how it is providing food for its own population and for the world. Built on an area that extends over 4,000 square meters, the pavilion is a dynamic and expansive structure with an ambitious form that surges skywards, combining unique engineering solutions and green technologies. Its simple, yet memorable facade is made from sustainable, ecologically-sound materials offer protection from the elements. Its voluminous hill becomes an element of the natural landscape, surrounded by organic wooden hoops that form the elegant "nose" of a sailing boat – a metaphor of Noah's Ark.

©Daniele Mascolo (courtesy of 2015 Expo Milano)

©Pietro Baroni (courtesy of 2015 Expo Milano)

©Pietro Baroni (courtesy of 2015 Expo Milano)

©Daniele Mascolo (courtesy of 2015 Expo Milano)

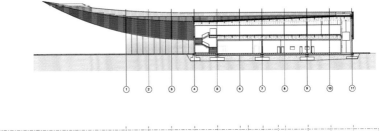

©Daniele Mascolo (courtesy of 2015 Expo Milano)

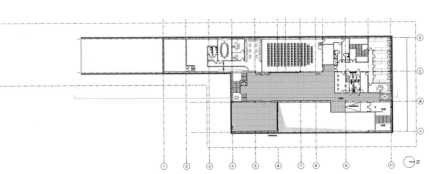

>>巴林国馆 _ Studio Anne Holtrop

绿色考古学

从迪尔蒙古代文明开始,巴林国就以其丰富而又独特的农耕历史而自豪。这一根深蒂固的农业传统是由于巴林岛上有大量甘甜的泉水,否则这地方将会非常贫瘠。2015米兰世博会上巴林国馆很好地诠释了这个国家农耕传统与文化之间的紧密关系。展馆的核心是向其丰富的农业遗产致敬,这其中包括十个各具特色的水果园。在长达六个月的世博会展会期间,每个水果园都会在不同时间段结不同的水果。该展馆也将展示具有数千年历史的文物,每一件历史文物都与其盘根错节的农耕传统和传说有关。据说,巴林国是传说中伊甸园的所在,也是一百万株棕榈树之乡。展馆由建筑师Anne Holtrop设计,景观建筑师是Anouk Vogel,展馆的设计是连续的巴林国果园景观与一系列封闭的展览空间相互贯穿,纵横交错。展馆由白色预制混凝土板建造,便于在世博会结束后拆卸运回巴林国重建成植物园。用于建造展馆的预制构件彼此之间的接缝清晰可见,意在说明在巴林国考古发现中其传统建筑固有的形状。

>> Bahrain _ Studio Anne Holtrop

Archaeologies of Green

Since the ancient civilization of Dilmun, Bahrain has boasted a rich and unique agrarian history. This deep-rooted heritage is underpinned by the plentiful sweet water springs which exist in this otherwise arid land. The Kingdom of Bahrain's pavilion at the 2015 Milan Expo presents an interpretation of the relationships that tie together the country's agrarian heritage and culture. The centerpiece of the pavilion pays homage to this rich heritage, which consists of 10 distinct fruit gardens, each of which will bear their fruit at different times during the six-month long Expo. The pavilion will also showcase historic artefacts that date back thousands of years, each of which is related to the deep-rooted agrarian traditions and the legends that surrounded Bahrain as the location of the Garden of Eden and the Land of One Million Palms. The pavilion was designed by the architect Anne Holtrop and landscape architect Anouk Vogel, and is conceived as a continuous landscape of Bahraini fruit gardens which intersect in a series of closed exhibition spaces. Built out of white prefabricated concrete panels, the pavilion will be moved to Bahrain at the end of the Expo and rebuilt to serve as a botanical garden. The prefabricated components of the buildings, visible through the seams that connect them to one another, refer to the inherent shapes found in the archaeology of Bahrain.

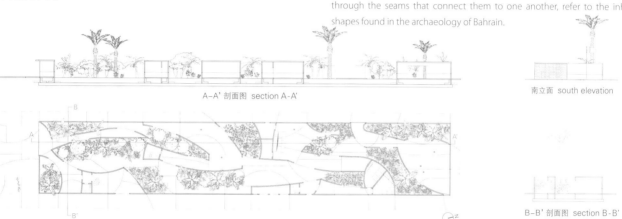

A-A' 剖面图 section A-A'

南立面 south elevation

B-B' 剖面图 section B-B'

©Daniele Mascolo (courtesy of 2015 Expo Milano)

©Armin Linke and Giulia Bruno (courtesy of Bahrain Authority for Culture & Antiquities)

>>阿联酋馆 _ Foster + Partners

精神食粮——塑造未来，共享未来

阿联酋馆探究了人类在给养地球方面，尤其是在如何解决土地、食物、能源和水资源这些错综复杂的问题方面所面临的真正挑战。该馆特别展示了阿联酋在这些领域所探索出的一些独创性解决办法。尽管这些解决办法的提出主要是基于当地的实际需要，但是由于不断上涨的需求和气候变化所带来的影响，世界上的许多地方将会经历和阿联酋一样的困难，所以说阿联酋所提出的这些独创性解决方案能使全球真正受益。通过分享知识和资源，阿联酋正真正地帮助塑造未来。该展馆由福斯特及合伙人事务所设计完成，一系列高高的波浪起伏的围墙成就其引人注目的外观。这些给人深刻印象的12m高的结构使人联想到了阿联酋古代居住区里狭窄的、自我遮蔽的人行道以及沙漠中宏伟开阔的沙丘。沙丘表面3D扫描技术使墙的设计具有真实的肌理和质感。蜿蜒曲折的墙体自然而然地引导着游客通过一系列有趣的空间，兴奋的同时了解了许多知识。用阿联酋本土的植物物种作为展馆的装饰与补充，给人带来了一种温馨、绿色的感觉。该展馆的设计既要考虑举办地米兰自然凉爽的天气，还要考虑阿联酋炎热的气候，因为在2015世博会结束后，该展馆将被拆卸并在阿联酋低碳城市马斯达尔市重建。这样做也是马斯达尔市可持续发展精神的体现。

>> UAE Pavilion _ Foster + Partners

Food for Thought – Shaping and Sharing the Future

The UAE Pavilion explores the very real challenges that arise in feeding the planet, particularly in the interwoven topics of land, food, energy and water. It also highlights some of the innovative solutions that the UAE has developed in these areas. Solutions have been arrived at based on a local need but they have a very real global benefit since many parts of the world will be experiencing the same difficulties as the Emirates due to spiralling demand and the effects of climate change. By sharing knowledge and resources, the UAE is truly helping to shape the future.

Designed by Foster + Partners, the striking form of the UAE Pavilion is created by a series of tall rippled walls. These impressive 12-meter structures evoke both the narrow self-shaded streets of the UAE's historic settlements and the magnificent open sand dunes of its deserts. 3D scans of dune surfaces have informed the wall design to create an authentic texture. These sinuous, curving shapes guide visitors through a range of intriguing spaces and exciting, informative experiences. Complementary landscaping based on native UAE species provides a welcoming green backdrop.

Engineered for the naturally cool weather of Milan and the sunny climes of the United Arab Emirates, the pavilion will be dismantled at the end of 2015 Expo Milano and reassembled in the UAE's low-carbon Masdar City, reflecting the principles embodied in Masdar's sustainability ethos.

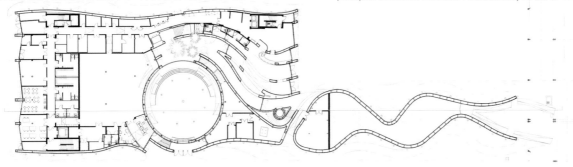

©Daniele Mascolo (courtesy of 2015 Expo Milano)

©Daniele Mascolo (courtesy of 2015 Expo Milano)

©Daniele Mascolo (courtesy of 2015 Expo Milano)

©Pietro Baroni (courtesy of 2015 Expo Milano)

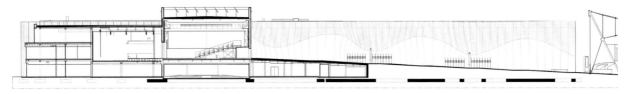

>>德国馆_Schmidhuber

灵感的田野

2015米兰世博会德国的口号是"积极参与",因此德国呈现给参观者的是"灵感的田野",一个富有活力、充满奇思妙想的展馆。

德国馆清晰地表明了这一主题,用实例生动地说明尊重自然对于未来人类的营养是多么的重要。展馆空间和其内容的紧密互动很好地体现了这一理念。"灵感的田野"体现在建筑上,展馆就像是缓慢升起的景观带,让人联想到德国特色的田野和草地。

德国馆最主要的设计元素之一是风格化的植物。从展馆一层开始,它们是"思想的幼苗",慢慢向上生长到达顶层表面,其树叶犹如大的华盖覆盖全馆,把室内和室外连为一体,使展览与建筑融为一体。

游客有两种不同的方式来参观这个"灵感的田野"。选择第一条线路,游客可以自由进入观景平台,闲庭散步,就像在公园一样逗留徘徊。在"思想的幼苗"冲破地面的地方,游客可以看到下面展览区的精彩内容,从而被激发起兴趣和好奇心。

选择第二条路线,游客可以到达展馆内部的专题展览区,这里展示了营养的不同来源——土壤、水、气候和生物的多样性以及城市中的食物生产和消费,不一而足。各种各样的主题展览、展品和展台都呈现了德国以一种让人吃惊的方式来解决人类面临的未来人类营养的挑战,鼓励游客积极参与到展馆活动之中。

>> Germany Pavilion_Schmidhuber

Fields of Ideas

Germany reveals itself at the 2015 Expo Milano as "Fields of Ideas" under the motto "Be active", as a vibrant, fertile "landscape" filled with ideas.

The Germany Pavilion clearly orients itself to this leitmotif – vividly illustrating just how important dealing respectfully with nature is to future human nutrition. The pavilion concept is characterized by close interaction between spatial and content presentation. The "Fields of Ideas" are reflected in the architecture – evoking Germany's distinctive field and meadow landscapes – in the form of a gently rising landscape level.

One of the pavilion's key design elements is stylized plants that grow as "Idea Seedlings" from the exhibition level to the surface, where they unfold into a large canopy of leaves. They connect indoor and outdoor spaces, exhibition and architecture.

Visitors can explore the "Fields of Ideas" in two different ways. Taking the first route, they can stroll along the freely accessible landscape level, which invites visitors to linger and enjoy as they would in a public park. Wherever the "Idea Seedlings" break through the surface, visitors are offered insight on highlights in the exhibition below – sparking interest and curiosity.

The second route leads visitors through the thematic exhibition inside the pavilion, which showcases the different sources of nutrition – Soil, Water, Climate and Biodiversity – all the way to food production and consumption in the urban world. A variety of thematic ambassadors, exhibits and stations present surprising approaches from Germany on addressing the challenge of future human nutrition. Visitors are invited to become active themselves.

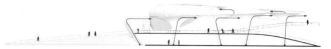

©B.Handke (courtesy of German Pavilion 2015 Expo Milano)

©Daniele Mascolo (courtesy of 2015 Expo Milano)

©Daniele Mascolo (courtesy of 2015 Expo Milano)

©B.Handke (courtesy of German Pavilion 2015 Expo Milano)

在2015米兰世博会上,意大利救助儿童协会也建了一个展馆。这个馆被定义为一个"村子"更合适,有家一样温馨的环境,让人感到宾至如归。就像NGO(非政府组织)通常设立的拯救儿童村一样,这里是一个理想的世界。

一个个房间没有墙,也没有其他的围挡,完全与周围的环境和其他的展馆融为一体。空间通透、开放,外部景观与室内空间连为一体,就像拯救儿童协会在许多国家所做的一样,周围环境是家庭和社区生活的重要组成部分。2015世博会拯救儿童馆的设计源于对木材和金属板等这些简单材料的研究,地板是用混凝土和泥土做的,目的是使人们想起遥远的故乡的温暖和色彩。

在拯救儿童馆里我们还可以发现"自产"这一主题:展馆周边所使用的面板是杉木和竹子做的,出自一个工作坊中参加"Civico Zero"项目的外国孩子之手。

室外景观的设计是为了欢迎、引导、娱乐以及教育公众:参观体验植物的同时了解关于它们的故事,那些古老起源的谷物,树木和一个小菜园,所有这些都是为了让参观者想起社区的土地分配,这对拯救儿童的项目来说非常重要。大人和小孩都可以通过所安装的一些互动设施来了解拯救儿童馆,既有趣又有教育意义。

展览从世博会核心议题开始,按照"营养、营养不良、突发事件"三个主要主题,介绍了拯救儿童协会这一非政府组织所做的工作。

Save the Children Village

On the occasion of 2015 Expo Milano, Save the Children Italia ONLUS presents a pavilion that could be better defined as a proper "village", a domestic and welcoming environment which designs the space of an ideal place of the world where the NGO usually operates.

A sequence of rooms without walls and barriers but in conversation with the surroundings and other pavilions. A space that is permeable and open, where the external landscape is an integral part of the internal world, such as the territory is a strong part of the life of families and communities in the countries where Save the Children works. Save the Children Village for Expo 2015 has been conceived starting from a

拯救儿童馆
Argot ou La Maison Mobile

research of simple materials as wood and metal sheet, together with a floor made by concrete and soil, with the aim to remind the warmth and colors of far away territories.

A village where we can also find the theme of "self-production": the perimeter panels made of fir wood and bamboos have been made through a workshop with foreign kids who are part of the program "Civico Zero".

The project of the outdoor landscape has been created to welcome, to guide, to entertain and to educate the public: vegetal experiences that bring stories with them, cereals of ancient origins, trees and a small vegetable garden, to recall the community allotments that are very important about Save the Children programs. Adults and kids discover the village through interactive installations which are funny and educational at the same time.

The exhibition tells about the work of the ONG starting from the Expo main topic, following three main themes: Nutrition/Malnutrition/Emergency.

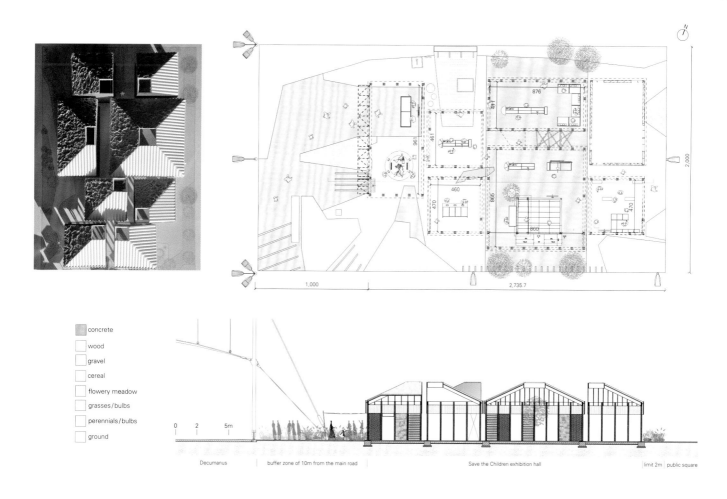

- concrete
- wood
- gravel
- cereal
- flowery meadow
- grasses/bulbs
- perennials/bulbs
- ground

Decumanus | buffer zone of 10m from the main road | Save the Children exhibition hall | limit 2m | public square

摄影师：©Delfino Sisto Legnani (courtesy of the architect) (except as noted)

慢食馆
Herzog & de Meuron

慢食运动组织和瑞士建筑设计公司赫尔佐格&德梅隆合作完成的2015米兰世博会慢食馆5月1日起开始对游客开放，5月19日慢食运动组织总裁卡洛·彼得里尼和瑞士建筑师赫尔佐格一起正式为慢食馆揭牌剪彩。

赫尔佐格&德梅隆最初参与了2015米兰世博会的整体规划，但是当他们意识到"主办方不会采取必要步骤说服参与国放弃根深蒂固的自我思维模式而是专注于各自国家对农业和粮食生产做出的实际贡献"时，他们曾不再考虑这一项目。

当慢食运动组织决定参加2015世博会，在这个国际平台上发出自己的声音的时候，按照赫尔佐格&德梅隆的要求建造一个符合2015世博会"给养地球，生命的能源"主题的富有创新性的世博会展馆就至关重要了。慢食运动组织与赫尔佐格&德梅隆的愿景一致，认为世博会应该注重内容而不是一些华而不实的、不可再次使用的结构，这样做只会让游客误解本次展览会的真正目的。因此，慢食运动组织决定与赫尔佐格&德梅隆合作共同建造一个符合建筑师最初整体规划的特别展馆。

2014年，慢食运动组织委托赫尔佐格&德梅隆来承担世博会本展馆的设计任务。他们答应后就立即开始与慢食运动组织一起设计慢食馆。赫尔佐格&德梅隆能接受设计请求令慢食运动组织非常开心，因为这就意味着能够看到他们重要的合作伙伴如何用建筑来诠释和实施

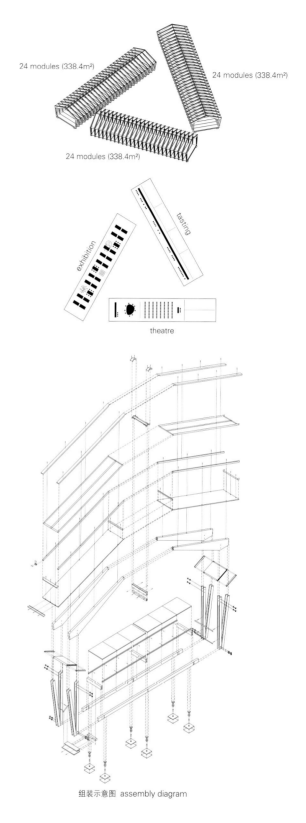

组装示意图 assembly diagram

Slow Food Pavilion

Slow Food and the Swiss architecture firm Herzog & de Meuron have collaborated on the construction of the Slow Food Pavilion for Expo 2015 in Milan, which was accessible from May 1 and officially inaugurated on May 19 in the presence of Slow Food president Carlo Petrini and the Swiss architect Jacques Herzog.

Herzog & de Meuron were originally involved in the imple-

该组织的主要议题和关注点。接下来就是与赫尔佐格&德梅隆非同寻常的合作,设计师们对慢食运动组织的理念、主题和风格都有非常精准的理解。

该展馆应该为游客展示农业和食品生态多样性的意义,探索各种生态多样性的产品,并使其意识到采用新型消费习惯的必要性。

建筑设计和管理方案基于一个简单的表格式布局,营造出食堂和市集的氛围。人们可以观看视觉陈述和阅读关键词,了解不同的消费习惯以及它们对我们地球的影响;可以与可持续农业拥护者和地方粮食生产商见面、交谈和讨论,了解关于可替代方法的知识;还可以通过嗅觉和味觉感受丰富的农业和食品生态多样性。

1 展览区 2 蔬菜园 3 查询处/书店/牛奶吧 4 品尝区 5 办公室 6 仓库 7 储物柜 8 准备室
9 酒窖 10 清洁室 11 当地废物处理区 12 奶酪准备室 13 售酒商店 14 收银处 15 奶酪品尝室
1. exhibition   2. vegetable garden   3. info/book shop/milk bar   4. theater of taste   5. office   6. ware house   7. locker   8. preparation room
9. wine cellar   10. cleaning room   11. local waste disposal   12. cheese preparation room   13. wine shop   14. cash register   15. cheese tasting room
一层 first floor

1 展览区 2 蔬菜园 3 奶酪品尝室 4 奶酪准备室
1. exhibition   2. vegetable garden   3. cheese tasting room   4. cheese preparation room
A–A' 剖面图

mentation of their master plan for 2015 Expo Milano, but abandoned the project when they realized that "the organizers would not undertake the necessary steps to convince the participating nations to give up on their conventional indulging in self-contemplation instead of focusing on their specific contribution to agriculture and food production."

When Slow Food decided to participate at 2015 Expo to make its voice heard in this international platform, it became important to take on Herzog & de Meuron's requirement to create an innovative Expo space that would be in harmony with the theme of 2015 Expo: Feeding the Planet. Energy for Life. Slow Food agrees with Herzog & de Meuron's vision of focusing on the content of the exposition rather than on pompous and unsustainable structures that would only distract visitors from the real purpose of the event. Slow Food was therefore certain that it would like to collaborate with Herzog & de Meuron on the implementation of a special pavilion that would stick to the architects' original master plan.

In 2014, Slow Food asked Herzog & de Meuron to take on the task of designing its space at Expo, and once they accepted, they started to work with Slow Food on the pavilion. It was extremely gratifying for Slow Food that the architects accepted, because it meant being able to see the organization's main topics and concerns be interpreted and implemented architectonically by such an important partner. What followed was an extraordinary collaboration with Herzog & de Meuron, because they have shown a very accurate comprehension of the Slow Food philosophy, themes, and style.

The pavilion should allows visitors to discover the significance of agricultural and food biodiversity, to explore the variety of the products that are protagonists of biodiversity, and to become aware of the need of adopting new consumption habits. The architectural and curatorial proposal is based on a simple layout on tables which creates an atmosphere of refectory and market. People can watch visual statements and read key texts about different consumption habits and their consequences for our planet, they can meet and discuss with exponents of sustainable agriculture and local food production to learn about alternative approaches, and they can smell and taste the richness of agricultural and food biodiversity.

用地面积: 3,272m² / 有效楼层面积: 1,170m² / 摄影师: ©Marco Jetti

>>48

EHDD

Marc L'Italien is a Design Principal at EHDD and one of the firm's most innovative thinkers and designers. For over 20 years he has designed public buildings with challenging technical requirements including museums, aquariums, etc. His deep expertise and involvement in all phases of exhibit-based design, as well as the regulatory issues involved with large urban re-use projects, are reflected in the creative and commercial success of his projects. His inventive work minimizes dependence on natural resources, focuses on public learning, and celebrates and protects wild animal species.

>>72

Open Architecture

It was founded by Li Hu[left] and Huang Wenjing[right] in New York City. They won WA Chinese Architecture Award and Chinese Architecture Media Award's Best Architecture Award in 2002 and established the Beijing office in 2006. Li Hu is a partner at Steven Holl Architects. He founded and led SHA's Beijing office as the partner-in-charge for the firm's many award winning projects in Asia. Huang Wenjing is a senior designer and associate at Pei Cobb Freed and Partners Architects. She is a visiting assistant professor at the University of Hong Kong and teaches part time at Tsinghua University in Beijing as well.

>>88

Ryuichi Ashizawa Architect & Associates

Ryuichi Ashizawa was born in Yokohama, Japan in 1971 and graduated from the Waseda University in 1994. He has worked for Tadao Ando Architect & Associates for 6 years before founding his own office Ryuichi Ashizawa Architects & Associates in 2001. He has been an Associate Professor at the University of Shiga Prefecture since 2013. He won various prizes at the JCD Award, Good Design Award, DFA Award and Energy Globe Awards.

CoBe Architecture
Since its inception in 2002, the agency CoBe placed the habitat and landscape in the heart of their reflections. Based on this multidisciplinary approach, they develop projects both across urban (creation of new districts and urban renewal), housing (construction, rehabilitation) and various scales of landscape (public spaces, parks and gardens, landscape studies).
Anxious to better respond to contemporary issues, the agency is committed to a voluntary approach, based on an environmental approach, a typological invention and control costs by seeking alternative and efficient solutions.

Mutabilis Paysage
Ronan Gallais[right], the co-founder of Mutabilis Paysage, was born in 1972, Rennes. He graduated in 2000 from the National School of Higher Studies in Landscape Architecture in Versailles and was an assistant professor from 2003 to 2008 at the National School of Higher Studies in Landscape Architecture in Brittany. In 2006, he was awarded the National Award for Landscape Architecture by the Minister of Culture, France. He is currently involved in the urban renewal of the Francois Mitterand Boulevard, Rennes, the urban renewal of Mulhouse City Center and the Development and Direction of the Spatial Planning of the region between Tarn and Toulouse.
Juliette Bailly-Maître[left], the co-founder of Mutabilis Paysage, was born in 1973, Fort de France Graduated in 1998 from the School of Beaux Arts and National School of Higher Studies in Landscape Architecture in Versailles. She was involved in teaching at the National School of Higher Studies in Landscape Architecture in Versailles and gave conferences on the projects. In 2006, she was awarded the National Award for Landscape Architecture by the Minister of Culture, France.

Lake|Flato Architects
One of the partners, Andrew Herdeg[left] has developed a body of work that reinforces essential human experiences at the intersection of architecture and landscape. Fostering the emotional, psychological and spiritual relationship between people and the natural environment, his thoughtful designs reflect a commitment to the study of place: climate, natural resources, cultural precedents, and historic building traditions.
Joseph Benjamin[right], the associate partner of the firm, has worked on projects across the country with a particular focus on nature centers and education facilities. Prior to his career in architecture, Joe worked in the construction industry for several years, which gave him a practical and pragmatic understanding of architecture. In his ten years with Lake|Flato, he has been committed to ensuring that sustainable issues – such as biomimicry, high-performance, material durability, water conservation and rainwater collection – are integrated from early design concepts through construction and preserved throughout the overall lifespan of the building.

>>60
Dissing+Weitling Architecture
It was founded in 1971 to continue the work started by Arne Jacobsen. Since then, the company has refined and developed Scandinavian design tradition with works that have set new standards in design and architecture. Over the years, they have built up a professional portfolio that includes some of the world's most spectacular bridges and distinctive international business headquarters, as well as new residential areas, historic building renovations, and interior and product design. Daniel V. Hayden[above] was born in 1961. He experienced a DIS studies (Danish Institute for Study Abroad) in 1982 and graduated from Rensselaer Polytechnic Institute, Troy, New York in 1984. After the graduation, he worked at several architectural offices in US and moved to Copenhagen in 1990. He joined Dissing+Weitling Architecture in 1992 and became a partner in 2002.

Kennedy & Violich Architecture, Ltd.
It was founded in 1990 by Sheila Kennedy[left] and J. Frano Violich[right]. Sheila Kennedy received her Bachelor's Degree in history, philosophy and literature from the College of Letters at Wesleyan University. She studied architecture at the Ecole National Supérieure des Beaux Arts in Paris and received the Master's degree of Architecture from the Graduate School of Design at Harvard University. She is described as an "insightful and original thinker who is designing new ways of working, learning, leading and innovating".
J. Frano Violich was born in San Francisco, California and grew up in a bi-lingual family with strong ties to Venezuela and South America. He received his Bachelor's degree in Architecture at the University of California at Berkeley and received the Master's degree of Architecture from the Graduate School of Design at Harvard University where he was graduated with Distinction, the School's highest academic honor.

>>28

C3, Issue 2015.9

All Rights Reserved. Authorized translation from the Korean-English language edition published by C3 Publishing Co., Seoul.

© 2016大连理工大学出版社
著作权合同登记06-2016年第73号

版权所有·侵权必究

图书在版编目(CIP)数据

能源意识与可持续公共空间：汉英对照 / 韩国C3出版公社编；安雪花等译. — 大连：大连理工大学出版社，2016.6

(C3建筑立场系列丛书)

书名原文：C3: Energy-conscious and Sustainable Public Spaces

ISBN 978-7-5685-0409-6

Ⅰ. ①能… Ⅱ. ①韩… ②安… Ⅲ. ①节能－建筑设计－汉、英 Ⅳ. ①TU2

中国版本图书馆CIP数据核字(2016)第136620号

出版发行：大连理工大学出版社
　　　　　（地址：大连市软件园路80号　邮编：116023）
印　　刷：上海锦良印刷厂
幅面尺寸：225mm×300mm
印　　张：11.75
出版时间：2016年6月第1版
印刷时间：2016年6月第1次印刷
出 版 人：金英伟
统　　筹：房　磊
责任编辑：杨　丹
封面设计：王志峰
责任校对：周小红
书　　号：978-7-5685-0409-6
定　　价：228.00元

发　行：0411-84708842
传　真：0411-84701466
E-mail：12282980@qq.com
URL：http://www.dutp.cn